T0211135

Communications
in Computer and Information Science 2005

Rationale

The CCIS series is devoted to the publication of proceedings of computer science conferences. Its aim is to efficiently disseminate original research results in informatics in printed and electronic form. While the focus is on publication of peer-reviewed full papers presenting mature work, inclusion of reviewed short papers reporting on work in progress is welcome, too. Besides globally relevant meetings with internationally representative program committees guaranteeing a strict peer-reviewing and paper selection process, conferences run by societies or of high regional or national relevance are also considered for publication.

Topics

The topical scope of CCIS spans the entire spectrum of informatics ranging from foundational topics in the theory of computing to information and communications science and technology and a broad variety of interdisciplinary application fields.

Information for Volume Editors and Authors

Publication in CCIS is free of charge. No royalties are paid, however, we offer registered conference participants temporary free access to the online version of the conference proceedings on SpringerLink (http://link.springer.com) by means of an http referrer from the conference website and/or a number of complimentary printed copies, as specified in the official acceptance email of the event.

CCIS proceedings can be published in time for distribution at conferences or as post-proceedings, and delivered in the form of printed books and/or electronically as USBs and/or e-content licenses for accessing proceedings at SpringerLink. Furthermore, CCIS proceedings are included in the CCIS electronic book series hosted in the SpringerLink digital library at http://link.springer.com/bookseries/7899. Conferences publishing in CCIS are allowed to use Online Conference Service (OCS) for managing the whole proceedings lifecycle (from submission and reviewing to preparing for publication) free of charge.

Publication process

The language of publication is exclusively English. Authors publishing in CCIS have to sign the Springer CCIS copyright transfer form, however, they are free to use their material published in CCIS for substantially changed, more elaborate subsequent publications elsewhere. For the preparation of the camera-ready papers/files, authors have to strictly adhere to the Springer CCIS Authors' Instructions and are strongly encouraged to use the CCIS LaTeX style files or templates.

Abstracting/Indexing

CCIS is abstracted/indexed in DBLP, Google Scholar, EI-Compendex, Mathematical Reviews, SCImago, Scopus. CCIS volumes are also submitted for the inclusion in ISI Proceedings.

How to start

To start the evaluation of your proposal for inclusion in the CCIS series, please send an e-mail to ccis@springer.com.

Enhong Chen · Yang Gao · Longbing Cao ·
Fu Xiao · Yiping Cui · Rong Gu · Li Wang ·
Laizhong Cui · Wanqi Yang
Editors

Big Data

11th CCF Conference, BigData 2023
Nanjing, China, September 8–10, 2023
Proceedings

 Springer

Editors
Enhong Chen
University of Science and Technology
of China
Hefei, China

Longbing Cao
Macquarie University
Sydney, NSW, Australia

Yiping Cui
Nanjing Normal University
Nanjing, China

Li Wang
Taiyuan University of Technology
Taiyuan, China

Wanqi Yang
Nanjing Normal University
Nanjing, China

Yang Gao
Nanjing University
Nanjing, China

Fu Xiao
Nanjing University of Posts
and Telecommunications
Nanjing, China

Rong Gu
Nanjing University
Nanjing, China

Laizhong Cui
Shenzhen University
Shenzhen, China

ISSN 1865-0929   ISSN 1865-0937 (electronic)
Communications in Computer and Information Science
ISBN 978-981-99-8978-2   ISBN 978-981-99-8979-9 (eBook)
https://doi.org/10.1007/978-981-99-8979-9

This Springer imprint is published by the registered company Springer Nature Singapore Pte Ltd.
The registered company address is: 152 Beach Road, #21-01/04 Gateway East, Singapore 189721, Singapore

Paper in this product is recyclable.

Preface

Welcome to the proceedings of the 11th CCF Big Data Conference (BigData 2023), which was held in Nanjing, China, from September 8 to 10, 2023. BigData 2023 was themed "Collaborating Computility and Model, Data Intelligence Leading the Future" to jointly discuss the opportunities and challenges faced by big data in the era of digital economy and large models. The aim of BigData 2023 was to provide a high-quality platform for researchers and practitioners from academia, industry, and government to share their research results, technical innovations, and applications in the field of big data.

The topics of the accepted papers include theories and methods of data science, algorithms, and applications of big data. The papers were all comprehensively double-blind reviewed and evaluated by three to four qualified and experienced reviewers from relevant research fields. The 14 full papers accepted for publication were selected from 69 submissions.

On behalf of the organizing committee, our thanks go to the keynote speakers for sharing their valuable insights with us and to the authors for contributing their work to this conference. We would like to express sincere thanks to China Computer Federation (CCF), CCF Expert Committee on Big Data, Nanjing University, Nanjing Normal University, Nanjing University of Posts and Telecommunications, Yangtze River Delta Information Intelligence Innovation Research Institute, Jiangsu Computer Society, and Jiangsu Association of Artificial Intelligence for their support and sponsorship. We would also like to express our deepest appreciation to the Technical Program Committee members, reviewers, session chairs, and volunteers for their strong support in the preparation of this conference.

Last but not least, we highly appreciate Springer publishing the proceedings of BigData 2023.

November 2023

Enhong Chen
Yang Gao
Longbing Cao

Organization

Honorary Chairs

Guojie Li — Institute of Computing Technology, Chinese Academy of Sciences, China

Hong Mei — Peking University, China

Steering Committee Chairs

Xiaoyong Du — Renmin University of China, China

Xueqi Cheng — Institute of Computing Technology, Chinese Academy of Sciences, China

Hai Jin — Huazhong University of Science and Technology, China

General Chairs

Jian Lv — Nanjing University, China

Ninghui Sun — Institute of Computing Technology, Chinese Academy of Sciences, China

Yike Guo — Hong Kong University of Science and Technology, China

Program Committee Chairs

Enhong Chen — University of Science and Technology of China, China

Yang Gao — Nanjing University, China

Longbing Cao — Macquarie University, Australia

Organizing Committee Chairs

Fu Xiao	Nanjing University of Posts and Telecommunications, China
Yiping Cui	Nanjing Normal University, China
Rong Gu	Nanjing University, China

Publication Chairs

Li Wang	Taiyuan University of Technology, China
Laizhong Cui	Shenzhen University, China
Wanqi Yang	Nanjing Normal University, China

Communication Chairs

Rui Mao	Shenzhen University, China
Yidong Li	Beijing Jiaotong University, China
Fei Teng	Southwest Jiaotong University, China
Gong Cheng	Nanjing University, China

Session Chairs

Junping Du	Beijing University of Posts and Telecommunications, China
Jianxin Li	Beijing University of Aeronautics and Astronautics, China
Yinghuan Shi	Nanjing University, China

Best Paper Awards Committee Chairs

Xiaoyang Wang	Fudan University, China
Peng Cui	Tsinghua University, China

Finance Chairs

Lixuan Shen	China Computer Federation, China
Rui Bao	Nanjing University, China

Sponsorship Chairs

Huaiming Song	Golaxy, China
Wujun Li	Nanjing University, China

Program Committee

Bin Zhao	Nanjing Normal University, China
Bin Zhou	National University of Defense Technology, China
Bo Jin	Third Research Institute of the Ministry of Public Security, China
Chang Tan	iFLYTEK Co., Ltd, China
Changsheng Li	Beijing Institute of Technology, China
Chuan Shi	Beijing University of Posts and Telecommunications, China
Defu Lian	University of Science and Technology of China, China
Denghao Ma	Meituan, China
Deqing Wang	Beijing University of Aeronautics and Astronautics, China
Di Wu	Sun Yat-sen University, China
Dian Shen	Southeast University, China
Dong Zhang	Inspur Electronic Information Industry Stock Price Co., Ltd, China
Fan Jiang	Jingdong Technology Group, China
Fei Liu	Hefei University of Technology, China
Fengyi Song	Nanjing Normal University, China
Fuyuan Cao	Shanxi University, China
Gang Xiong	China Academy of Sciences, China
Ge Song	Nanjing Normal University, China
Gong Cheng	Nanjing University, China
Guangyan Zhang	Tsinghua University, China
Guiyun Zhang	Tianjin Normal University, China
Guoliang He	Wuhan University, China
Guoxian Yu	Southwest University, China
Haitao Zhang	Beijing University of Posts and Telecommunications, China
Hanhua Chen	Huazhong University of Science and Technology, China
Hao Liu	Hong Kong University of Science and Technology (Guangzhou), China

Haoran Cai	Huawei Technologies Co., Ltd, China
Hongzhi Wang	Harbin Institute of Technology, China
Hu Ding	University of Science and Technology of China, China
Hua Dai	Nanjing University of Posts and Telecommunications, China
Huawei Shen	China Academy of Sciences, China
Hudong Li	Beijing Jiaotong University, China
Huihui Wang	Nanjing University of Technology, China
Jianquan Ouyang	Xiangtan University, China
Jianxin Li	Beijing University of Aeronautics and Astronautics, China
Jianzong Wang	Ping An Technology (Shenzhen) Co., Ltd, China
Jiawei Luo	Hunan University, China
Jie Wang	University of Science and Technology of China, China
Jieyue He	Southeast University, China
Jing Huo	Nanjing University, China
Jing Zhang	Renmin University of China, China
Ju Fan	Renmin University of China, China
Jun Long	Central South University, China
Junchi Yan	Shanghai Jiao Tong University, China
Ke Li	Beijing Union University, China
Laizhong Cui	Shenzhen University, China
Lei Qi	Southeast University, China
Lei Wang	Institute of Computing, Chinese Academy of Sciences, China
Li Kong	Nanjing Normal University, China
Liang Bai	Shanxi University, China
LiKe Xin	Nanjing Normal University, China
Limin Xiao	Beijing University of Aeronautics and Astronautics, China
Lin Shang	Nanjing University, China
Ling Lin	Yili Normal University, China
Miao Cai	Hohai University, China
Minling Zhang	Southeast University, China
Peizhen Peng	Nanjing Normal University, China
Qi Liu	University of Science and Technology of China, China
Qinghua Zhang	Chongqing University of Posts and Telecommunications, China
Qiulan Huang	Brookhaven National Laboratory, USA
Rong Gu	Nanjing University, China

Ruixuan Li	Huazhong University of Science and Technology, China
Shangdong Yang	Nanjing University of Posts and Telecommunications, China
Shijun Liu	Shandong University, China
Shuai Wang	Southeast University, China
Tao Jia	Southwest University, China
Tao Zhou	Beijing Venustech Information Security Technology Co., Ltd, China
Tieying Zhang	Alibaba Silicon Valley Lab, China
Wanqi Yang	Nanjing Normal University, China
Weisheng Xie	China Telecom Tianyi E-commerce Co., Ltd, China
Wenbin Li	Nanjing University, China
Wujun Li	Nanjing University, China
Xiang Zhang	Southeast University, China
Xiang Zhao	National University of Defense Technology, China
Xiangnan He	University of Science and Technology of China, China
Xiaofeng Gao	Shanghai Jiao Tong University, China
Xiaojun Chen	Shenzhen University, China
Xiaolong Jin	Institute of Computing Technology, Chinese Academy of Sciences, China
Xuan Zhou	East China Normal University, China
Yan Yang	Southwest Jiaotong University, China
Yanhong Guo	Dalian University of Technology, China
Yi Du	Computer Network Information Center, Chinese Academy of Sciences, China
Yilei Lu	White Sea Technology, China
Yinghuan Shi	Nanjing University, China
Yong Zhang	Shenzhen Institute of Advanced Technology, Chinese Academy of Sciences, China
Yongbin Qin	Guizhou University, China
Yongxin Tong	Beijing University of Aeronautics and Astronautics, China
Yubao Liu	Sun Yat-sen University, China
Yunquan Zhang	Chinese Academy of Sciences, China
Zhaohong Deng	Jiangnan University, China
Zhaokang Wang	Nanjing University of Aeronautics and Astronautics, China
Zhichao Zheng	Nanjing Normal University, China
Zhihong Shen	China Academy of Sciences, China

Zhipeng Gao	Beijing University of Posts and Telecommunications, China
Zhirong Shen	Xiamen University, China
Zhiyong Peng	Wuhan University, China
Zhongbao Zhang	Beijing University of Posts and Telecommunications, China
Zhonghong Ou	Beijing University of Posts and Telecommunications, China
Zili Zhang	Southwest University, China

Contents

Long-Term and Short-Term Perception in Knowledge Tracing

Zihang Chen[1], Mengxiao Zhu[1]($\boxtimes$), Fei Wang[2], Shuanghong Shen[2],
Zhenya Huang[2], and Qi Liu[2]

[1] Anhui Province Key Laboratory of Science Education and Communication,
Institute of Advanced Technology and School of Humanities and Social Sciences,
University of Science and Technology of China, Hefei, China
`czh1999@mail.ustc.edu.cn, mxzhu@ustc.edu.cn`
[2] Anhui Province Key Laboratory of Big Data Analysis and Application, School of
Data Science and School of Computer Science and Technology, University of Science
and Technology of China, Hefei, China
`{wf314159,shshen}@mail.ustc.edu.cn, {huangzhy,qiliuql}@ustc.edu.cn`

Abstract. Knowledge Tracing (KT) is a fundamental task in contemporary intelligent educational systems, which tracks the knowledge states of the learners based on their response sequences. KT is crucial for the effectiveness of computer-assisted intelligent educational systems, such as intelligent tutoring systems and educational resource recommendation systems. In recent years, KT models benefited from the deep learning approaches and improved dramatically compared with the traditional probabilistic approaches. However, deep learning based KT models also have significant limitations. For example, Recurrent Neural Network (RNN)-based KT models can not capture dependencies in long-term sequences effectively, and attention mechanism-based KT models rely on the positional encoding to perceive sequential information, which may disrupt the semantics of the original embeddings. Moreover, the data sparsity question remains a big challenge in existing KT models. This study is based on the observation that KT shows a stronger sequential dependence in the long term than in the short term. In this paper, we propose a novel KT model called "Long-term and Short-term perception in knowledge tracing (LSKT)", which uses multilayer perceptrons and attention mechanism to capture the long-term and short-term dependencies, and employs 2PL-IRT based embedding to alleviate the data sparsity question. Extensive experiments on multiple datasets demonstrate the effectiveness of our proposed LSKT model.

Keywords: Knowledge Tracing · Educational Data Mining · Intelligent Education · Deep Learning

1 Introduction

In the past three decades, online learning platforms have experienced rapid development, leading to the rise of intelligent education. One fundamental task in

E. Chen et al. (Eds.): BigData 2023, CCIS 2005, pp. 1–15, 2023.
https://doi.org/10.1007/978-981-99-8979-9_1

intelligent education is to track the evolving knowledge states of learners in order to provide personalized learning resources and plan learners' learning paths, which is known as Knowledge Tracing (KT) [1]. Current KT models utilize learners' historical question-answering records to infer their abilities to accurately answer new questions, which dynamically assess the learners' knowledge states. Using this approach, intelligent education tasks such as providing appropriately challenging questions and arranging personalized learning paths are achieved.

In recent years, with the rapid advancement of deep learning, the latest studies on KT are based on deep learning models. These studies employ sequence models in deep learning, such as Recurrent Neural Networks (RNNs) [2] and Attention Mechanisms [3], to model learners' question-answering sequences. Due to the significant progress in deep learning, these deep learning-based knowledge tracing models have achieved outstanding results.

However, despite the accomplishments of previous methods, there are still inherent limitations. For instance, KT models based on RNNs can not effectively capture the long-term sequential dependencies. As for the attention-based KT models, the original attention mechanisms are not sensitive to sequential dependencies in input sequences, requiring additional processing, such as adding positional embeddings, to make the model aware of the sequence information. Existing attention-based knowledge tracing models add the learner's answer record embeddings and positional embeddings together as inputs to the attention module, which may disrupt the underlying semantic of the original answer record embeddings [4] and lead to a decline in model performance.

The recent developments in Multi-Layer Perceptrons (MLPs), such as MLP-mixers [5] and gMLP [6], offer a new approach for capturing sequential dependencies in learners' question-answering records. This approach employs simple MLPs in the sequence dimension, replacing RNNs and attention mechanisms. By doing so, it effectively captures long-term sequential dependencies and avoids the influence of positional embeddings.

On the other hand, in many user modeling tasks, researchers separate user interaction sequences into long-term and short-term to model them separately. For instance, in recommendation systems, researchers distinguish between users' long-term and short-term interests and model them differently [7,8]. Inspired by these methods, we have applied a similar idea to KT. From a long-term perspective (the entire answer sequence of learners), their knowledge state constantly changes, and these changing knowledge states also reflect their abilities. From a short-term perspective (several most recent answer records), a given learner's knowledge state is relatively stable, allowing for better diagnosis of the current level of knowledge mastery. Moreover, we think that long-term answer sequences exhibit strong sequential dependencies, while the sequential dependencies in short-term answer sequences are weaker. To treat short-term answers as sequences would introduce unnecessary computations and potentially reduce the model's generalization ability.

Additionally, an inevitable challenge in the KT task is the sparsity of data: many questions are assigned to a small number of students, making it difficult for current KT models to effectively represent the questions. Inspired by the two-parameter logistic item response theory (2PL-IRT) [9] model, which uses two parameters, difficulty and discrimination, to distinguish different questions with surprising results, we think that utilizing two trainable scalars can effectively represent distinct questions while mitigating over-parameterization and overfitting challenges stemming from data sparsity.

Based on these perspectives, we propose a model called Long-term and Short-term perception in Knowledge Tracing (LSKT). It utilizes MLPs to perceive the long-term sequential dependencies in learners' answer sequences and employs attention mechanisms without any positional embeddings to perceive short-term dependencies. In addition, we have also proposed an embedding method based on 2PL-IRT that effectively mitigates the data sparsity in KT. To validate the effectiveness of the LSKT approach, we conduct extensive experiments and compare it with multiple baseline models on various datasets. Additionally, we conduct ablation experiments to further analyze and evaluate the contributions of long-term perception and short-term perception. In summary, our paper presents the following contributions:

- We propose a novel model that utilizes MLPs to extract sequential dependencies in KT. This model avoids the additional embedding influence when using attention mechanisms and overcomes the limitations of RNNs in capturing long-term sequential dependencies.
- We differentiate between long-term and short-term dependencies in KT and employ different methods for modeling each.
- Inspired by the 2PL-IRT model, we propose a novel embedding method for KT that effectively addresses the issue of data sparsity in KT and alleviates the resulting questions of over-parameterization and overfitting.
- We conduct extensive experiments on multiple datasets, demonstrating the superior performance of the LSKT model and the effectiveness of its individual modules.

2   Related Work

2.1   Knowledge Tracing

Existing KT models can be broadly divided into two categories: traditional probabilistic models and deep learning-based methods. In some special cases, traditional approaches may approximate the performance of deep learning-based methods. However, in the vast majority of cases, deep learning-based methods are much more powerful. The most fundamental issue in KT is capturing the sequential dependencies in learners' answer sequences, i.e., perceiving the constantly changing knowledge states of learners and the trends in these knowledge state changes.

The most classic KT model is Bayesian Knowledge Tracing (BKT) [1], which uses binary variables to represent a learner's knowledge state as either mastered or not mastered. BKT employs a hidden Markov model to sequentially model learners' response behaviors, treating whether the learner has mastered the knowledge as a hidden state and whether the learner can answer correctly as an observable state, capturing the sequential dependencies through transition state matrices.

On the other hand, deep learning-based KT models represent a learner's knowledge mastery level using a high-dimensional vector, providing richer information compared to BKT's binary representation. The first deep learning-based KT model is Deep Knowledge Tracing (DKT) [10]. DKT utilizes recurrent neural networks to model learners' answer sequence, with the hidden layer of the recurrent neural network considered as the representation of the learner's knowledge state. As learners' responses update and their knowledge state changes, DKT updates the learner's knowledge state through representations of questions and learner responses. Several extensions of DKT, such as DKT+ [11] and IEKT [12], have achieved promising results.

With the introduction and development of attention mechanisms, attention-based KT models have also achieved outstanding performance. The first attention based KT model is Self-Attentive Knowledge Tracing (SAKT) [13], which calculates the weights of the past answer behaviors in the current sequence dependency using attention mechanisms and obtains more complex dependency relationships by stacking multiple attention layers. Subsequently, the more advanced attention-based KT model SAINT [14] was proposed. It employs the Transformer architecture to capture sequential dependencies. The Transformer architecture consists of an encoder for modeling question information and a decoder for modeling learner responses. With a more sophisticated architecture, SAINT outperforms SAKT in terms of performance.

In addition to the above methods, there are many other deep learning-based KT approaches. For example, Dynamic Key-Value Memory Networks (DKVMN) [15] model knowledge tracing using dynamic key-value pairs; Convolutional Knowledge Tracing (CKT) [16] employs convolutional neural networks for modeling; Graph-based Interaction Knowledge Tracing (GIKT) [17] utilizes graph convolutional neural networks for modeling, and so on. However, despite the significant progress achieved by current methods, there are still certain inherent limitations, as discussed in Sect. 1.

2.2   Recent Advances in MLP

The Multilayer Perceptron (MLP) is a basic architecture of artificial neural networks and one of the most common and earliest deep learning models. In the early era of machine learning, MLP achieved excellent performance. However, due to its lack of capabilities in handling sequential data, natural language, and images, MLP has been replaced by RNNs, convolutional neural networks (CNNs), attention mechanisms, and other models.

In recent years, researchers have found that simple modifications to the MLP architecture can lead to remarkable results in the aforementioned tasks. Specifically, the original MLP only projects the input data along a single dimension, which limits its ability to handle sequential data, images, and other complex data types. However, by transposing the input data, it becomes possible to project along another dimension, enabling the handling of sequences, images, and similar data. By performing multiple alternating projections along different dimensions, complex dependency relationships can be captured.

The first work that sparked a resurgence of MLP was MLP-Mixers [5] in the computer vision field. It modified the MLP architecture using the aforementioned approach and replaced the attention module in the Transformer architecture. The results showed state-of-the-art performance in image classification benchmarks. Subsequently, MLP-based approaches emerged in other domains as well. For example, Moi-Mixer [18] utilized MLP for sequence recommendation, and MTS-Mixers [19] employed MLP for time series forecasting. However, the application of MLP in KT has not been explored.

3   Question Definition

In this section, we provide a formal definition of the **Knowledge Tracing** task, a fundamental challenge within intelligent educational systems. The core objective of this task is to predict a learner's ability to provide accurate responses to future questions based on their historical interactions.

3.1   Concepts and Data Representation

Within the context of an intelligent educational system, we consider a collection of question items denoted as $Q = \{q_1, q_2, q_3, \ldots, q_{numq}\}$, and a set of associated knowledge concepts represented as $C = \{c_1, c_2, c_3, \ldots, c_{numc}\}$. Each question corresponds to a specific knowledge concept. In cases where a question relates to multiple knowledge concepts, these concepts are treated as a composite knowledge entity. Learners' responses to questions are binary in nature, with 1 indicating a correct answer and 0 indicating an incorrect answer.

3.2   Interaction Record Representation

The interaction history of a learner within the intelligent educational system can be succinctly captured as a sequence of tuples: $\{(q_1, c_1, r_1), (q_2, c_2, r_2), \ldots, (q_n, c_n, r_n)\}$. Each tuple (q_t, c_t, r_t) constitutes a fundamental unit of interaction, where q_t signifies the presented question, c_t denotes the corresponding knowledge concept, and r_t signifies the correctness of the learner's response.

3.3   Objective of Knowledge Tracing

At the heart of the knowledge tracing question lies the task of estimating a learner's state of knowledge at a specific step t. This involves the prediction of a learner's ability to accurately address questions associated with q_{t+1} and c_{t+1} in the subsequent step $t + 1$. In simpler terms, given a learner's interaction history up to step t, the goal is to provide insights into the learner's preparedness for upcoming questions concerning a novel question q_{t+1} and its corresponding knowledge concept c_{t+1}.

4   Method

In this section, we introduce the architecture of the proposed model, as shown in the Fig. 1. It consists of three layers: 2PL-IRT based embedding layer, long-term and short-term perception layer, and learner's response prediction layer. The detailed description of these three layers is as follows.

4.1   2PL-IRT Based Embedding Layer

In KT tasks, we inevitably confront the challenge of data sparsity. In an intelligent educational system, the number of questions often significantly outstrips the amount of learners, with many questions being assigned to a small number of learners. For example, in ASSIST2009 [20], there were 4,217 learners, while the number of questions reached 26,688. Therefore, in some previous studies, researchers merely used KC to index questions without distinguishing different question IDs. However, this approach is evidently flawed, as different questions sharing the same KC may vary in terms of difficulty and discrimination. Learners with the same knowledge state may exhibit different performances when faced with different questions that share the same KC.

Inspired by the simple yet powerful 2PL-IRT model in psychometrics, we propose a new approach to address this issue. We differentiate different questions with the same KC through two trainable scalar parameters. More specifically, we represent the question and interaction at step t in the following way:

$$
\begin{aligned}
\mathbf{x}_t &= \mathbf{c}_t + \mathbf{c}'_t \cdot Repeat(\mathbf{i}_t, d); \\
\mathbf{y}_t &= \mathbf{x}_t + \mathbf{a}_t,
\end{aligned}
\tag{1}
$$

where $\mathbf{x}_t$ and $\mathbf{y}_t$ are the representation of question and interaction at step t. $Repeat(\cdot, d)$ signifies expanding the tensor's dimension to d by repeating and d is the embedding size of KC. $\mathbf{c}_t$, $\mathbf{c}'_t$, $\mathbf{i}_t$ and $\mathbf{a}_t$ are latent representation of the original input, which are obtained as following:

$$
\begin{aligned}
\mathbf{c}_t &= \mathbf{W}_c \cdot \mathbf{e}_{c_t}; \\
\mathbf{c}'_t &= \mathbf{W}_{c'} \cdot \mathbf{e}_{c_t}; \\
\mathbf{i}_t &= \mathbf{W}_i \cdot \mathbf{e}_{q_t}; \\
\mathbf{a}_t &= \mathbf{W}_a \cdot \mathbf{r}_t,
\end{aligned}
\tag{2}
$$

where $\mathbf{c}_t$ and $\mathbf{c}'_t$ are the latent representation of KC contained in the question at step t, $\mathbf{i}_t$ is the IRT factor symbolizes difficulty and discrimination, $\mathbf{a}_t$ is the representation of learner's response at step t. $\mathbf{e}_{c_t}$ denotes the one-hot encoding of the KC contained in the question at step t. $\mathbf{e}_{q_t}$ denotes the one-hot encoding of the question id at step t, and $\mathbf{r}_t$ indicates the one-hot encoding of the learner's correctness of answer at step t. $\mathbf{W_c} \in \mathbb{R}^{d \times numc}$, $\mathbf{W_{c'}} \in \mathbb{R}^{d \times numc}$, $\mathbf{W_i} \in \mathbb{R}^{2 \times numq}$ and $\mathbf{W_a} \in \mathbb{R}^{d \times 2}$ are learnable linear transformation operations. $numc$ is the total number of KCs and $numq$ is the total number of questions in the dataset.

4.2   Long-Term and Short-Term Perception Layer

Next, we introduce the Long-term and Short-term Perception Layer used to perceive the learner's knowledge state. We introduce the long-term perception module and the short-term perception module separately.

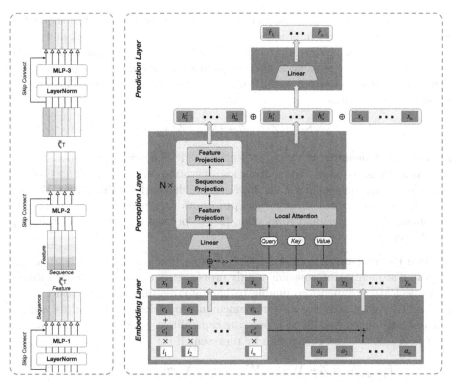

Fig. 1. The architecture of the LSKT model. The image on the left depicts the details of the long-term perception module.

MLP for Long-Term Perception. As mentioned before, we employ MLP to perceive the long-term dependencies in the learner's interaction sequence to circumvent the limitations of attention mechanisms and RNNs, which is depicted in the left part of Fig. 1. Specifically, we concatenate the learner's question records and interaction records embeddings as the input to the long-term perception module:

$$\mathbf{I}_l = \mathbf{W}_1 \cdot (\mathbf{X} \oplus (\mathbf{Y} \gg 1)), \tag{3}$$

where $\mathbf{X} = \{\mathbf{x}_1, \mathbf{x}_2, \ldots, \mathbf{x}_n\}$ and $\mathbf{Y} = \{\mathbf{y}_1, \mathbf{y}_2, \ldots, \mathbf{y}_n\}$ are the output of the embedding layer, which are the latent representation of the questions and interactions. $\oplus$ is the concatenation symbol and $\gg$ represents shifting one step to the right which is used to prevent label leakage. Then $\mathbf{X} \oplus (\mathbf{Y} \gg 1) = \{\mathbf{x_1} \oplus \mathbf{0}_d, \mathbf{x_2} \oplus \mathbf{y_1}, \ldots, \mathbf{x_n} \oplus \mathbf{y_{n-1}}\}$, where $\mathbf{0}_d$ represents a vector of dimension d

and each element is 0. $\mathbf{W_1} \in \mathbb{R}^{d \times 2d}$ is a learnable linear transformation operation to ensure $\mathbf{I}_l$ has the same dimensions as $\mathbf{X}$ and $\mathbf{Y}$.

The long-term perception module is a stacked structure composed of multiple sub-modules. Each sub-module consists of two parts: sequence projection and channel projection.

Sequence Projection. The sequence projection aims to learn the sequential dependencies in the user interaction series. Assuming that the input matrix for sequence projection is $\mathbf{Z} \in \mathbb{R}^{d \times n}$, where d is the embedding size and n is the length of interaction sequences, then sequence projection linearly projects the input, producing an output matrix of the same shape $(d \times n)$. This can be specifically represented as:

$$S_{proj}(\mathbf{Z}) = \mathbf{Z} + (\mathbf{W_2} \cdot \mathbf{Z}^\top)^\top, \tag{4}$$

where $S_{proj}(\cdot)$ denotes sequence projection and $\top$ denotes transpose. $\mathbf{W_2} \in \mathbb{R}^{n \times n}$ is a learnable linear transformation operation and n is the length of the learner's interaction sequence. The lower triangular parts of the matrix $\mathbf{W_2}$ are set to 0 and not altered through backpropagation to prevent label leakage. In this phase, We use Residual connection [21] to prevent degradation of the deep neural network.

Feature Projection. Although sequential projection has effectively learned and integrated sequential information, it only processes sequential information within the same feature dimension, unable to handle cross-dimensional feature information. Therefore, we have also implemented projection on the feature dimension and introduced non-linear processing capabilities. Specifically, assuming the input is $\dot{\mathbf{Z}} \in \mathbb{R}^{d \times n}$, the output can be represented as:

$$F_{proj}(\dot{\mathbf{Z}}) = \dot{\mathbf{Z}} + \mathbf{W_4} \cdot \sigma(\mathbf{W_3} \cdot LayerNorm(\dot{\mathbf{Z}}) + \mathbf{b_1}) + \mathbf{b_2}, \tag{5}$$

where $F_{proj}(\cdot)$ denotes feature projection and $LayerNorm$ is Layer normalization [22]. $\mathbf{W_3}, \mathbf{W_4}, \mathbf{b_1}, \mathbf{b_2}$ are trainable parameters and $\mathbf{W_3}, \mathbf{W_4} \in \mathbb{R}^{d \times d}, \mathbf{b_1}, \mathbf{b_2} \in \mathbb{R}^{d \times 1}$. σ is a nonlinear activation function like ReLU.

In the sub-module of Long-term Perception Module, we have adopted one sequence projection and two feature projections, arranged in the order of $F_{proj} \rightarrow S_{proj} \rightarrow F_{proj}$. Therefore, the learner's knowledge state as perceived by the Long-term Perception Module would be:

$$\mathbf{H}^l = F_{proj}(S_{proj}(F_{proj}(\mathbf{I}_l))). \tag{6}$$

And $\mathbf{h}_t^l \in \mathbf{H}^l = \{\mathbf{h}_1^l, \mathbf{h}_2^l, \ldots, \mathbf{h}_n^l\}$ denotes the knowledge state perceived long-term perception module at step t.

Attention for Short-Term Perception. As mentioned earlier, we use simple dot product attention to perceive short-term dependencies in learner interaction sequences. Specifically, the learner's knowledge state as perceived by the Short-term Perception Module would be:

$$\mathbf{H}^s = \{\mathbf{h}_1^s, \mathbf{h}_2^s, \ldots, \mathbf{h}_n^s\};$$
$$\mathbf{h}_t^s = Attention(Q = \mathbf{x}_t, K = \{\mathbf{x}_{t-T}, \ldots, \mathbf{x}_{t-1}\}, V = \{\mathbf{y}_{t-T}, \ldots, \mathbf{y}_{t-1}\}), \tag{7}$$

where *Attention* is Scaled Dot-Product Attention [23] and T represents the size of the short-term perception window that we have set.

4.3   Response Prediction Layer

The purpose of the last layer of the model is to predict the learner's response to the current question. In the prediction layer, the knowledge states extracted by the long-term and short-term perception modules and the question embeddings are combined as input, which is then further processed by a fully connected neural network. The processed output is transformed into a predictive probability through a sigmoid activation function. This probability value represents the likelihood of the learner answering the current question correctly. To be more specific, the probability of the learner answering the current question correctly at step t is:

$$\hat{r}_t = sigmoid(\mathbf{w} \cdot \sigma(\mathbf{W}_6 \cdot \sigma(\mathbf{W}_5 \cdot (\mathbf{h}_t^l \oplus \mathbf{h}_t^s \oplus \mathbf{x}_t) + \mathbf{b}_3) + \mathbf{b}_4) + b), \tag{8}$$

where $\mathbf{W}_5$, $\mathbf{W}_6$, $\mathbf{b}_3$, $\mathbf{b}_4$, $\mathbf{w}$ and $\mathbf{b}$ are trainable parameters and $\mathbf{W}_5 \in \mathbb{R}^{d \times 3d}$, $\mathbf{W}_6 \in \mathbb{R}^{d \times d}$, $\mathbf{w} \in \mathbb{R}^{1 \times d}$, $\mathbf{b}_3$, $\mathbf{b}_4 \in \mathbb{R}^{d \times 1}$, b is a scalar. In LSKT, all learnable parameters are trained in an end-to-end fashion by minimizing the binary cross-entropy loss over all learner responses, which is:

$$\mathcal{L}(\theta) = -\sum_i \sum_t (r_t^i \log(\hat{r}_t^i) + (1 - r_t^i) \log(1 - \hat{r}_t^i)), \tag{9}$$

where θ denotes all parameters of LSKT, more details of settings will be specified in the section experiments.

5   Experiments

This section provides a comprehensive overview of various experiments conducted on multiple real-world datasets.

5.1   Datasets

In this paper, to evaluate the performance of LSKT, we conduct our experiments on four widely used datasets. The detailed descriptions are provided below::

ASSISTments2009 (AS2009): This dataset [20] consists of math exercises collected from the free online tutoring platform ASSISTments during the 2009–2010 school year. The dataset comprises 346,860 interactions involving 4,217 learners and 26,688 questions. It has served as the standard benchmark for knowledge tracing methods over the past decade and has been widely used.

NIPS34: This dataset [24] is derived from Tasks 3 & 4 of the NeurIPS 2020 Education Challenge. It collects learners' answers to multiple-choice diagnostic math questions and is gathered from the Eedi platform. For each question, we have selected the leaf nodes from the subject tree as its Knowledge Components (KCs). The final dataset consists of 1,382,727 interactions, 948 questions, and 57 KCs.

Algebra2005 (AL2005): This dataset is from the KDD Cup 2010 EDM Challenge [25] and it contains responses of 13–14 year old learners to Algebra questions. It provides detailed step-level learner responses. The dataset comprises 809,694 interactions involving 574 learners, 210,710 questions, and 112 KCs.

Bridge2006 (BR2006): This dataset is also from the KDD Cup 2010 EDM Challenge [25]. It includes 3,679,199 interactions, involving 1,146 learners, 207,856 questions, and 493 KCs.

5.2 Baselines

We compared LSKT with seven KT baselines, the details of which were compared as follows:

DKT [10] uses recurrent neural networks to model learners' learning behavior. In our implementation, we utilized LSTM.

DKT+ [11] is an improvement over the original DKT model aimed at enhancing the consistency of knowledge tracking predictions. It achieves this goal by introducing regularization terms corresponding to reconstruction and volatility into the original DKT model's loss function.

DKVMN [15] is a model that predicts students' knowledge mastery level directly by exploiting the relationships between latent knowledge components stored in a static memory matrix key and based on the dynamic memory matrix value.

SAKT [13] applies attention mechanisms to capture the connection between the learner and the interaction of the question.

SAINT [14] uses the transformer architecture to address KT placing the questions and learner responses in the encoder and decoder for processing, respectively.

ATKT [26] incorporates an attention-based LSTM model. This approach applies adversarial perturbations to the initial learner interaction sequences, aiming to mitigate the overfitting and insufficient generalization issues of the KT model.

IEKT [12] evaluates learners' knowledge status through the learner Cognition and Knowledge Acquisition Assessment module.

simpleKT [27] is a powerful and simple baseline method for handling KT tasks, modeling question-specific variations based on the Rasch model, and using ordinary dot-product attention functions to extract temporal perceptual information embedded in student learning interactions.

5.3   Experimental Setup

Similar to previous studies on KT models, we randomly selected 20% of the learner interaction sequences for model evaluation and performed standard 5-fold cross-validation on the remaining 80% of the sequences. Adam optimizer is used to optimize the model parameters, and the learning rate is set to 0.001. The maximum training period is set to 200, and an early stop mechanism is employed to expedite the model's training. The embedding dimension is 256, and the maximum length of learner interaction sequences is 200. The encoder consists of 4 stacked layers, and the size of the short-term perception window is set to 8. The model is implemented in PyTorch and trains using an NVIDIA A100 40GB PCIe. Like most existing methods, the primary evaluation metric is AUC, while ACC is a secondary metric. To ensure standardized experiments, we utilize pyKT [28], a Python-based benchmarking platform, to conduct our experiments.

5.4   Experimental Results

Our overall prediction performance (i.e., AUC and ACC) is presented in Table 1. In the table, the best result for each column is highlighted in boldface. From Table 1, we have observed the following results:

- As shown in Table 1, our proposed LSKT model outperforms all of the 7 baselines and achieves the best performance. Furthermore, it has demonstrated excellent performance on the AS2009 dataset, surpassing the best baseline by a significant margin of 1.32% in terms of AUC.
- We also observed that methods solely relying on attention, such as SAKT and SAINT, yielded poorer results. This could be attributed to the negative impact of positional encoding.

Table 1. The overall performance. LSKT outperforms all baselines on all datasets.

Model	AS2009		NIPS34		AL2005		BR2006	
	AUC	ACC	AUC	ACC	AUC	ACC	AUC	ACC
DKT	0.8229	0.7669	0.7718	0.7043	0.9194	0.8684	0.7951	0.8537
DKT+	0.8248	0.7675	0.7716	0.7055	0.9186	0.8685	0.7978	0.8547
DKVMN	0.8158	0.7600	0.7706	0.7035	0.9158	0.8648	0.7953	0.8535
SAKT	0.7546	0.7236	0.7469	0.6835	0.9060	0.8593	0.7611	0.8472
SAINT	0.7598	0.7316	0.7833	0.7144	0.8680	0.8400	0.7383	0.8382
ATKT	0.8332	0.7697	0.7762	0.7068	0.9223	0.8700	0.8079	0.8563
IEKT	0.7802	0.7348	0.8004	0.7289	0.8330	0.8202	0.8068	0.8533
SimpleKT	0.8384	0.7729	0.8023	0.7301	0.9249	0.8717	0.8111	0.8556
LSKT	**0.8494**	**0.7802**	**0.8028**	**0.7308**	**0.9277**	**0.8738**	**0.8162**	**0.8567**

5.5   Ablation Study

To further investigate the roles of different modules in LSKT, we conducted additional ablation experiments. We designed four comparative scenarios to enhance our understanding:

MovS removes the short-term perception module, meaning that we solely rely on the MLP to perceive the learner's knowledge state.

MovL removes the long-term perception module, meaning that we solely rely on the attention to perceive the learner's knowledge state.

MovIRT removes the IRT-based embedding module, meaning that we perform high-dimensional embedding for the question IDs, similar to KC (Knowledge Component), and added the results to the embedding of KCs to obtain the final embedding.

MovQ does not perform any embedding for the question ID, meaning that we only use KCs to represent a question.

The results of ablation experiments on four datasets are shown in Fig. 2. From the data analysis in Fig. 2, we can clearly observe that the overall performance of LSKT will decline regardless of which module is removed from the model. This strongly proves that all these modules play a key role in enhancing the performance of LSKT. We can also clearly see that removing the long-term perception module has a much greater impact on performance than removing the short-term perception module. This further underscores the importance of capturing long-term sequential dependencies in KT tasks, and the effectiveness of our long-term perception module in capturing these dependencies. In addition, we found that rather than using high-dimensional embeddings to represent question IDs, not representing question IDs actually yields better results in most datasets. This highlights the significant challenge posed by the data sparsity of question IDs in KT tasks. However, the 2PL-IRT based embedding method proposed in this paper can mitigate this issue to a certain extent.

5.6   Hyper-parameters Analysis

In this section, we will discuss how to adjust the basic hyperparameters of LSKT model to optimize its performance. We investigate the impact of the number of layers (N) and the size of the short-term perception window (T) on LSKT's ability to predict learner performance in the ASSIST2009 dataset.

The results in Fig. 3 show the impact of hyperparameters on the model performance. Particularly, when T is set to 8 and N is set to 4, the best results are achieved. On the other hand, we also notice that even in the case of N = 1, our model still demonstrates impressive performance. This means that our model can achieve excellent results even with a very low parameter count.

6   Conclusion and Future Work

In this paper, we proposed a novel KT model named LSKT, which effectively integrates the long-term and short-term dependencies in KT tasks using MLP

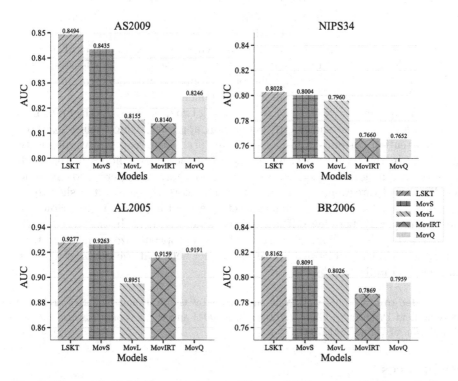

Fig. 2. The result of ablation experiments. We use AUC as our evaluation criterion.

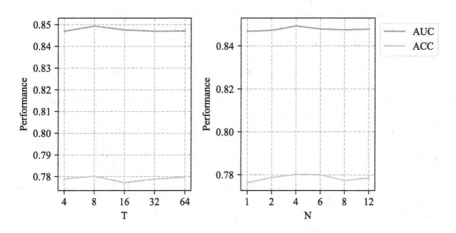

Fig. 3. The impact of hyperparameters.

and attention mechanisms respectively, and enables us to better capture the dynamics in KT. Additionally, we introduced an embedding method based on 2PL-IRT, which enables the better handeling of the data sparsity issue in KT tasks. Through comprehensive comparisons with seven commonly used baseline models on four real-world datasets, we provided evidence of the effectiveness of LSKT. Moreover, by conducting ablation experiments, we examined and proved the usefulness and contribution of each component in LSKT.

For future studies, we consider the following two directions. As the first direction, we focus on the short-term perception approach, which relies solely on simple attention currently and has a similar mechanism as cognitive diagnostic questions. In the future work, we intend to introduce techniques in the existing cognitive diagnostic models to enhance our short-term perception module. As the second direction, we plan to further explore the size of the short-term perception window, which is currently set as a fixed hyperparameter. However, the size may vary between different datasets and learners. In the future work, we plan to enable the dynamic learning of the appropriate size of the short-term perception window, which will make our model better fit the needs of different datasets and individual learners.

Acknowledgment. This work was supported by the National Key Research and Development Program of China (2021YFF0901004) and the National Natural Science Foundation of China (62177044).

References

1. Corbett, A.T., Anderson, J.R.: Knowledge tracing: modeling the acquisition of procedural knowledge. User Model. User-Adapt. Interact. **4**, 253–278 (1994)
2. Elman, J.L.: Finding structure in time. Cogn. Sci. **14**(2), 179–211 (1990)
3. Bahdanau, D., Cho, K., Bengio, Y.: Neural machine translation by jointly learning to align and translate. arXiv preprint arXiv:1409.0473 (2014)
4. Zheng, J., Ramasinghe, S., Lucey, S.: Rethinking positional encoding. arXiv preprint arXiv:2107.02561 (2021)
5. Tolstikhin, I.O., et al.: MLP-mixer: an all-MLP architecture for vision. In: Advances in Neural Information Processing Systems, vol. 34, pp. 24261–24272 (2021)
6. Liu, H., et al.: Pay attention to MLPs. In: Advances in Neural Information Processing Systems, vol. 34, pp. 9204–9215 (2021)
7. Tang, J., Wang, K.: Personalized top-n sequential recommendation via convolutional sequence embedding. In: Proceedings of the Eleventh ACM International Conference on Web Search and Data Mining, pp. 565–573 (2018)
8. Li, Y., et al.: Lightweight self-attentive sequential recommendation. In: Proceedings of the 30th ACM International Conference on Information & Knowledge Management, pp. 967–977 (2021)
9. Embretson, S.E., Reise, S.P.: Item Response Theory. Psychology Press (2013)
10. Piech, C., et al.: Deep knowledge tracing. In: Cortes, C., et al. (eds.) Advances in Neural Information Processing Systems, vol. 28. Curran Associates Inc. (2015). https://proceedings.neurips.cc/paper_files/paper/2015/file/bac9162b47c56fc8a4d2a519803d51b3-Paper.pdf

11. Yeung, C.-K., Yeung, D.-Y.: Addressing two problems in deep knowledge tracing via prediction-consistent regularization. In: Proceedings of the Fifth Annual ACM Conference on Learning at Scale, pp. 1–10 (2018)
12. Long, T., et al.: Tracing knowledge state with individual cognition and acquisition estimation. In: Proceedings of the 44th International ACM SIGIR Conference on Research and Development in Information Retrieval, pp. 173–182 (2021)
13. Pandey, S., Karypis, G.: A self-attentive model for knowledge tracing. In: 12th International Conference on Educational Data Mining, EDM 2019, pp. 384–389. International Educational Data Mining Society (2019)
14. Choi, Y., et al.: Towards an appropriate query, key, and value computation for knowledge tracing. In: Proceedings of the Seventh ACM Conference on Learning@ Scale, pp. 341–344 (2020)
15. Zhang, J., et al.: Dynamic key-value memory networks for knowledge tracing. In: Proceedings of the 26th International Conference on World Wide Web, pp. 765–774 (2017)
16. Shen, S., et al.: Convolutional knowledge tracing: modeling individualization in student learning process. In: Proceedings of the 43rd International ACM SIGIR Conference on Research and Development in Information Retrieval, pp. 1857–1860 (2020)
17. Yang, Y., et al.: GIKT: a graph-based interaction model for knowledge tracing. In: Hutter, F., Kersting, K., Lijffijt, J., Valera, I. (eds.) ECML PKDD 2020. LNCS (LNAI), vol. 12457, pp. 299–315. Springer, Cham (2021). https://doi.org/10.1007/978-3-030-67658-2_18
18. Lee, H., et al.: MOI-mixer: improving MLP-mixer with multi order interactions in sequential recommendation. arXiv preprint arXiv:2108.07505 (2021)
19. Li, Z., et al.: MTS-mixers: multivariate time series forecasting via factorized temporal and channel mixing. arXiv preprint arXiv:2302.04501 (2023)
20. Feng, M., Heffernan, N., Koedinger, K.: Addressing the assessment challenge with an online system that tutors as it assesses. User Model. User-Adap. Inter. **19**, 243–266 (2009)
21. He, K., et al.: Deep residual learning for image recognition. In: Proceedings of the IEEE Conference on Computer Vision and Pattern Recognition, pp. 770–778 (2016)
22. Ba, J.L., Kiros, J.R., Hinton, G.E.: Layer normalization. arXiv preprint arXiv:1607.06450 (2016)
23. Vaswani, A., et al.: Attention is all you need. In: Advances in Neural Information Processing Systems, vol. 30 (2017)
24. Wang, Z., et al.: Diagnostic questions: the NeurIPS 2020 education challenge. arXiv preprint arXiv:2007.12061 (2020)
25. Stamper, J., et al.: Algebra I 2005–2006 and bridge to algebra 2006–2007. In: Development Data Sets from KDD Cup 2010 Educational Data Mining Challenge (2010)
26. Guo, X., et al.: Enhancing knowledge tracing via adversarial training. In: Proceedings of the 29th ACM International Conference on Multimedia, pp. 367–375 (2021)
27. Liu, Z., et al.: simpleKT: a simple but tough-to-beat baseline for knowledge tracing. In: The Eleventh International Conference on Learning Representations (2023)
28. Liu, Z., et al.: pyKT: a python library to benchmark deep learning based knowledge tracing models. In: Thirty-sixth Conference on Neural Information Processing Systems Datasets and Benchmarks Track (2022)

A Transfer Learning Enhanced Decomposition-Based Hybrid Framework for Forecasting Multiple Time-Series

Yonghou He, Li Tao, and Zili Zhang$^{(\boxtimes)}$

School of Computer and Information Science, Southwest University, Chongqing, China
yhhe523@email.swu.edu.cn, {tli,zhangzl}@swu.edu.cn

Abstract. Time-series forecasting is a challenging task that requires high accuracy and efficiency. Hybrid models that combine decomposition algorithms with multiple individual models have demonstrated promising results for forecasting performance. However, these models also face the issues of high computational cost and time consumption when dealing with multiple time-series. To address these issues, this paper proposes a novel framework that integrates multivariate variational mode decomposition (MVMD) with auto-regressive integrated moving average (ARIMA) using transfer learning (TL), i.e. MVMD-ARIMA-TL. The framework decomposes multiple time series into sub-sequence groups with joint or common frequencies, which facilitates the transfer learning among similar sub-sequences by saving the paring process of source and target domain. The framework is evaluated on 5 real-world datasets from various domains such as energy consumption, network traffic, and solar radiation. The framework is compared with the conventional self-built MVMD-hybrid framework in terms of ARIMA model fitting time and normalized root mean square error (NRMSE) for forecasting accuracy. The results demonstrate that the proposed framework outperforms the conventional self-built framework by generating enhanced hybrid models with less model fitting time with the same NRMSE in most cases. This paper contributes to the literature by introducing a novel decomposition-based hybrid forecasting framework with transfer learning for multiple time-series that demand per hybrid model per time-series by addressing the issues of computing resource scarcity and high time consumption.

Keywords: computing resource scarcity · decomposition-based hybrid model · multiple time series forecasting · multivariate variational mode decomposition · transfer learning

1 Introduction

Time-series forecasting is the task of predicting the future values of a series of data points based on their past and present values. Multiple time-series forecasting is a more challenging task that involves predicting multiple interrelated

E. Chen et al. (Eds.): BigData 2023, CCIS 2005, pp. 16–31, 2023.
https://doi.org/10.1007/978-981-99-8979-9_2

series simultaneously. It has many applications in various domains such as finance [10], power industry [9], and transportation [11]. With the increasing demand to develop tools for forecasting futuristic events, time-series models that estimate future data trends based on historical information, such as individual and hybrid models, have been widely developed and adopted. Hybrid models are a combination of multiple individual models such as statistical models, machine learning models, and deep learning models. Hybrid models have prevailed in the past decades as they have been tested with better performance than those individual models to employ multi-methods to acquire the final results [15].

However, most of the existing hybrid models face the issues of high computational cost and time consumption when dealing with multiple time-series from different data owners such as smart electricity meters. In this condition, each data owner demands per hybrid model per time-series locally but lack computing resource themselves. However, conventional self-built hybrid models need to fit each sub-sequence with a suitable method from scratch for each data owner, which requires heavy computing locally and an amount of time. These issues arise when each data owner has multiple time-series of the same nature, such as multiple time-series representing different aspects or dimensions of a system or phenomenon, but they can not or will not share data and lack computing resources. Modeling multiple time-series collaboratively to improve the forecasting performance requires intensive computing and frequent communication among the data owners, which could be insatiable in data owner that lack computing resources and has privacy concern.

In this paper, we introduce and evaluate a novel framework that integrates multivariate variational mode decomposition (MVMD) with auto-regressive integrated moving average (ARIMA) network models using transfer learning (TL), i.e. MVMD-ARIMA-TL. This framework leverages the advantages of MVMD [13] to decompose multiple time-series into sub-sequence groups with joint or common frequencies, which facilitates the transfer of model parameters among similar sub-sequences by saving the clustering process. After decomposition for original time-seires, the ARIMA model is used for sub-sequences because it is quite suitable for simple time-series like these sub-sequences with a single principle frequency to save computing resources. By using transfer learning, this framework allows each data owner to build their decomposition-based hybrid model locally and collaboratively, without sharing their raw data or compromising their privacy. To conclude, we focus on two questions: 1) Can the decomposition-based hybrid model benefit from transfer learning? 2) Can simple statistical model ARIMA benefit from transfer learning? Focus on these two questions, our main contributions to the literature can be summarized as follows:

- **A novel decomposition-based hybrid forecasting framework with transfer learning is proposed to address the issues of high computational cost and time consumption of hybrid models for multiple time-series.**
- **Five real-world datasets from various domains such as energy consumption, network traffic, and solar radiation are used to validate**

the effect of transfer learning in decomposition-based hybrid time-series models.
- A finding that simple statistical models like ARIMA still can benefit from transfer learning is revealed by utilizing the common frequencies from decomposition by the MVMD, which could save the paring process of source and target domain in transfer learning.

The rest of the paper is organized as follows: Sect. 2 presents the methodology of our proposed framework and the strategy of transfer learning for time-series among various data owners. Section 3 shows the data analysis and experimental results of our proposed method on 5 real-world datasets and discusses the main findings, implications, and limitations of this study. Section 4 concludes the paper with a summary and some future work directions.

2   Method

This section describes the methods we used to test our hypothesis that transfer learning can improve the performance and efficiency of decomposition-based hybrid models for multiple time-series forecasting. We used 5 real-world datasets from various domains such as energy consumption, network traffic, and solar radiation. We applied the MVMD to decompose multiple time-series into sub-sequence groups with joint or common frequencies. We then built ARIMA models for each sub-sequence using transfer learning. We evaluated our framework by comparing it with the conventional self-built MVMD-hybrid framework in terms of ARIMA model fitting time and normalized root mean square error (NRMSE).

2.1   Datasets

We used 5 publicly available datasets from different domains: 1) Solar power: a subset of solar power generation from micro-generation units located in Évora city (Portugal) [8]; 2) Abilene: a subset of network traffic in the American Research and Education Network (Abilene) [18]; 3) GHI: a subset of solar radiation dataset of global horizontal irradiance from automated solar stations in Pakistan [16]; 4) Wind speed: a subset of wind speed dataset from automated solar stations in Pakistan [16]; 5) PV: a subset of Photovoltaic energy generation dataset in small businesses and residential households in southern Germany [6]. Table 1 summarizes the details of each dataset. We assumed that each data owner has only one time-series. We preprocessed the data by applying a clear sky model [1] to the solar power and PV dataset and removing night-time data points from the GHI datasets, which aim to normalize the data and remove the effects of weather and other external factors on solar power generation [8,14].

2.2   The Framework of Proposed Method

This subsection presents the framework of the proposed method for enhancing the performance and efficiency of decomposition-based hybrid models for multiple time-series forecasting through the integration of transfer learning. The

Table 1. Summary of datasets used.

Datasets	Unit/Station/Node	Data size	Time range		Time granularity	Train size	Test size (last day)	Preprocess
			Start date	End date				
Solar power	V8, V10, V11, V25, V40	9032	2011/2/1 1:00	2013/3/6 22:00	Hourly	9022	10	clear sky model [1]
Abilene	OD_4-4, OD_4-10, OD_7-6, OD_9-5, OD_9-9, OD_10-4, OD_11-10, OD_11-11, OD_12-4, OD_12-7	48096	2004/3/1 0:00	2004/9/10 23:55	5 min	47808	288	N/A
GHI	Lahore	3842	2017/3/12 11:10	2017/4/30 14:00	10 min	3760	82	night-time excluded
	Bahawalpur	23102	2015/9/18 1:40			23020		
	Islamabad	28470	2016/4/18 0:50			28388		
	Khuzdar	42738	2015/9/23 1:30			42656		
	Quetta	43108	2015/9/18 1:40			43026		
	Karachi	55274	2015/4/23 1:10			55192		
	Hyderabad	55352	2015/4/22 1:20			55270		
	Peshawar	56208	2015/4/11 1:00			56126		
	Multan	67420	2014/10/21 1:30			67338		
Wind speed	Lahore	7133	2017/3/12 11:20	2017/5/1 0:00	10 min	6988	145	N/A
	Bahawalpur	45305	2016/6/20 9:20			45160		
	Islamabad	54465	2016/4/17 18:40			54320		
	Hyderabad	84431	2015/9/22 16:20			84286		
	Karachi	84431	2015/9/22 16:20			84286		
	Khuzdar	84431	2015/9/22 16:20			84286		
	Multan	84431	2015/9/22 16:20			84286		
	Peshawar	84431	2015/9/22 16:20			84286		
	Quetta	84431	2015/9/22 16:20			84286		
PV	residential3	4480	2016/2/29 7:00	2017/3/6 15:00	Hourly	4470	10	clear sky model [1]
	residential6	5614	2015/10/25 8:00			5604		
	industrial1_1	5626	2015/10/24 6:00			5616		
	industrial1_2	5626	2015/10/25 6:00			5616		
	industrial3_facade	5714	2015/10/16 6:00			5704		
	industrial3_roof	5714	2015/10/16 6:00			5704		
	residential4	5768	2015/10/11 7:00			5758		
	residential1	7808	2015/5/21 18:00			7798		

framework, MVMD-hybrid-TL (see the example in Fig. 1), comprises three primary stages: decomposition of multiple time-series using the MVMD, construction of sub-models for each sub-sequence utilizing ARIMA with transfer learning, and reconstruction to generate forecasts for each time-series.

In the first stage, we employ the MVMD to decompose multiple time-series into sub-sequence groups characterized by joint or common frequencies. This algorithm enables the decomposition of multiple time-series to provide the similarity basis of each time series for transfer learning.

In the second stage, we construct sub-models for each sub-sequence by using ARIMA with transfer learning. This approach allows us to leverage knowledge from similar sub-sequences of multiple sources to enhance the forecasting performance and efficiency of target models.

In the third stage, we reconstruct the forecasts generated by each sub-model to produce the final forecasts for each time-series. We evaluate our framework by comparing it with the conventional self-built MVMD-hybrid framework in terms of model fitting time and NRMSE.

2.3   Baselines

In this subsection, we describe the baselines that we use to evaluate our framework for multiple time-series forecasting. We explain how we use different methods to obtain the start parameters for ARIMA models, i.e. Hannan Rissanen (HR) method [2], innovations maximum likelihood estimation (Innovations) method [3] and state-space method (Kalman) [4] for ARIMA models. We take these methods as the top line since they are based on the computation of data itself, which could be heavy computing and time-consuming when the time series is long. We also use no computation (Default set as zeros) for initialization as the bottom line for comparison in these two models since the transfer start parameters (TSPs) from other data owners are the same as Default in zero computing while other methods demand certain computing resources. See their complexity in Table 2.

Table 2. The time and space complexity of comparative methods. Note: N is the length of time series, k is the dimension of the state vector, p is the number of AR parameters, and q is the number of MA parameters.

Method	Time Complexity	Space Complexity
Innovations	$O(N^2)$	$O(N)$
Kalman	$O(k^2 N + k^3)$	$O(k^2 N + k^3)$
HR	$O((p+q)N)$	$O(N)$
Default	$O(p+q)$	$O(N)$
TSP	$O(p+q)$	$O(N)$

2.4   Proposed Transfer Learning Strategy

In this subsection, we describe the transfer effect of the proposed MVMD-hybrid-TL framework. These effects aim to improve the performance and efficiency of decomposition-based hybrid models for multiple time-series forecasting by transferring model parameters among similar sub-sequences and collaborating with other data owners. The overall working of this framework is depicted in Fig. 1 and its pseudo-code is shown in Algorithm 1. To be specific, the transfer effect leverages the knowledge representation generated by training models on

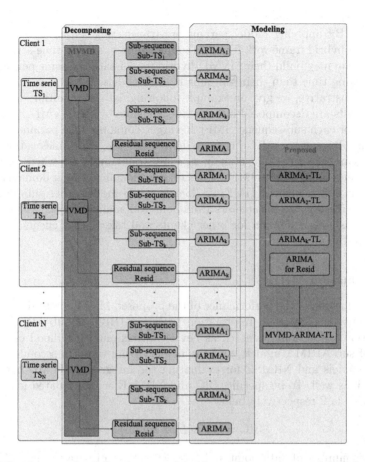

Fig. 1. The framework of the decomposition-based forecasting model with transfer learning (taking MVMD-ARIMA-TL as an example). Note: same color blocks represent similar time series/models; double-arrow lines represent communications/transmissions one way or both cross data owners.

Algorithm 1. MVMD-hybrid-TL (taking MVMD-ARIMA-TL as an example)

Input: multiple time series from multiple data owners (clients), $\{V\}_1^C$.

1: $\{\text{IMF}_{k,c}\}_{1,1}^{K,C} = MVMD(\{V\}_1^C)$ ▷ Decompose the multiple time series into sub-sequences.

2: **for** $k = 1 : K$ **do** ▷ This loop can be done parallel or successively.

3:   $\widehat{\text{IMF}}_{k,c} = ARIMA(\text{IMF}_{k,c})\_TSP(\text{IMF}_{k,C \neq c})$ ▷ ARIMA with start parameter from others clients $C \neq c$.

4: **end for**

5: $\{\widehat{V}\}_1^C = \{\sum_{k=1}^{K} \widehat{\text{IMF}}_{k,c} + \widehat{\text{Err}_c}\}_1^C$ ▷ Reconstruct the final forecasting results of one client by its sub-sequences forecasting results.

Output: forecasting results $\{\widehat{V}\}_1^C$.

---

one time-series and transfers it to enhance the forecasting performance on other time-series. By applying model parameters transferring among sub-models in the MVMD-hybrid framework from the source domain, similar time-series in the target domain may build their MVMD-hybrid model with the same performance but less fitting time than their self-built hybrid models.

To be illustrative, we give an example following: let's say multiple time series from C clients are decomposed into K sub-sequences by the MVMD, 1) parallel modeling, for each sub-sequence IMF_k in target domain c take parameters from same k-th sub-sequence IMF_k in source domains $C \neq c$, which means this target domain c can take any other domain as source domain $C \neq c$ when iteratively modeling each sub-sequence IMF_k; 2) successive modeling, target domain c could model some sub-sequences $\{IMF_k\}_1^{k_1}$ locally while model other sub-sequences $\{IMF_k\}_{k_1}^K$ utilizing the parameters from source domains $C \neq c$, which means each client is source domain for other clients and takes other clients as source domains in the meantime.

2.5   Evaluation Metrics

For the evaluation of the performance of the proposed MVMD-hybrid-TL framework, we use the Bayesian Information Criterion (BIC) of sub-ARIMA models, function evaluations (denoted as Feval) in maximum likelihood estimation (MLE) of sub-ARIMA models, and the model fitting time (in seconds) of sub-ARIMA models and NRMSE for evaluating the integrated MVMD-hybrid-TL framework as well. To be specific, BIC and NRMSE are formulated in Formula 1 and Formula 2 respectively below:

$$BIC = -2\ln(\hat{L}) + k\ln(T), \tag{1}$$

where T is number of data point and k is the number of parameters in ARIMA model; $\hat{L}$ is the maximized value of the likelihood function of the ARIMA model M, i.e. $\hat{L} = p(x|\hat{\theta}, M)$, where $\hat{\theta}$ are the parameter values that maximize the likelihood function, x is the observed data.

$$NRMSE = \frac{\sqrt{\frac{1}{T}\sum_{t=1}^{T}(\hat{\mathbf{V}} - \mathbf{V})^2}}{(\sum_{t=1}^{T}\mathbf{V})/T}, \tag{2}$$

where V is the original time-series, and $\hat{V}$ is the predicted value of time-series.

Since hybrid models with less BIC in sub-ARIMA models for sub-sequences do not forecast with less error always, we test hybrid models with metrics other than NRMSE, the sum of the model fitting time and Feval of sub-ARIMA models for each IMF in one time-series, and the maximum of model fitting time and function evaluations of MVMD-ARIMA for one time-series. To be specific, "time (sum)" and "Feval (sum)" evaluate the performance of the proposed framework for the successive decomposition-based hybrid model while "time (max)" and "Feval (max)" evaluate the performance of the proposed framework for the parallel decomposition-based hybrid model.

3   Results and Discussion

In this section, we present and analyze the results of our experiments on 5 real-world datasets from various domains. We compare our proposed framework with the conventional self-built MVMD-hybrid framework in terms of model fitting time and NRMSE. We also examine the effect of different methods for obtaining the start parameters for ARIMA models. We use different metrics and boxplots to evaluate the performance and efficiency of our framework.

3.1   Experimental Settings

To be specific, we assume each data owner has only one time-series since we use the ARIMA model for all time-series in transfer experiments. To be full proof of the efficiency of our framework, we traverse all time-series in each dataset by taking one data owner (one time-series) as the domain source and other data owners as the target domain in turn (only longer to equal-length or shorter time-series). For the decomposition of multiple time-series, the optimal number of modes (sub-sequences) is determined by the rule of least modes to make no mixing principle frequencies in any one IMF. For sub-ARIMA models, we set the order of the ARIMA model as the best order of the target model obtained by stepwise search based on BIC. To be consistent, the optimization methods used in the MLE of ARIMA models are all the same in one sub-sequence for different start parameters. To avoid convergence failure, multiple candidate methods are used in the optimization, which includes Limited-memory Broyden-Fletcher-Goldfarb-Shanno (L-BFGS) [5], modified Powell [12] and Nelder-Mead (NM) [17], and set with the same error tolerance and infinite iteration. All experiments are carried out in a Ubuntu platform: Intel(R) Xeon(R) CPU E5-2620 v4 @ 2.10 GHz 16 core CPU, 64G RAM.

3.2   Comparison of Time-Series and Its Sub-sequences

This subsection presents an example (dataset solar power) of the MVMD for multiple time-series, which aims to show the common frequencies of these time-series and explain how the MVMD could be used as a similarity measure. From Fig. 2a, it can be seen that when multiple time-series that represent the same kind of nature (i.e. these time-series are all about solar power generation in one place), they would show common patterns or characteristics among them (i.e. common frequencies in this study). From Fig. 2b to Fig. 2k, it can be seen that the commonality of frequencies among multiple time-series is a feasible property that could be used as asimilarity measure to facilitate the transfer learning and save the pairing process for the source and target domain since the MVMD would align the IMFs with common frequencies.

3.3   Comparison of Sub-ARIMA Models

This section presents the improvement of the MVMD-ARIMA approach with transfer learning against the self-build models, i.e. comparison for ARIMA with

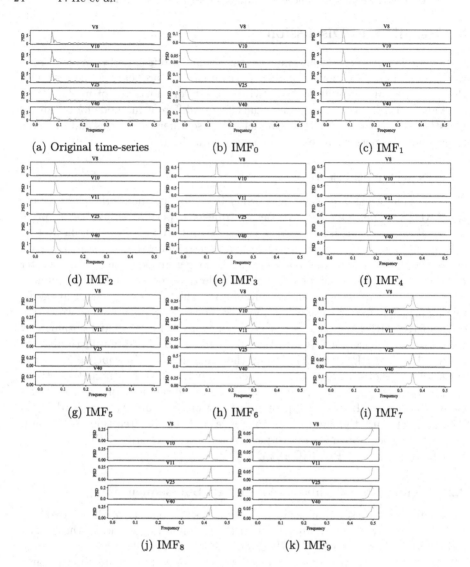

Fig. 2. The periodogram of the original time series (taking dataset solar power as the example) and its sub-sequences after decomposition by the MVMD.

TSP from the model of source sub-sequences, start parameters by HR, Innovations and Kalman, and Default setting start parameters with all zeros except parameter σ is set as one. From Figs. 3, 4, 5, 6 and 7, it can be seen that with TSPs from similar sub-models of time-series from other data owners, the sub-ARIMA model would have the same performance but less fitting time and/or fewer function evaluations in the optimization of MLE of ARIMA than self-built models.

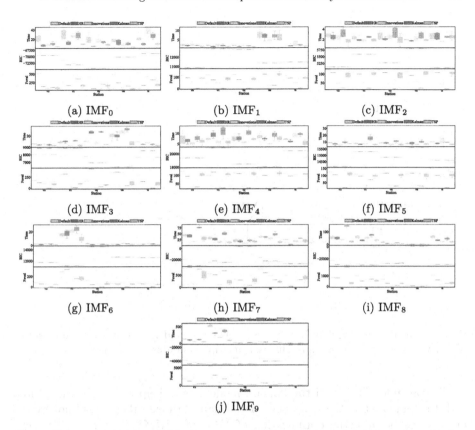

(a) IMF$_0$ (b) IMF$_1$ (c) IMF$_2$

(d) IMF$_3$ (e) IMF$_4$ (f) IMF$_5$

(g) IMF$_6$ (h) IMF$_7$ (i) IMF$_8$

(j) IMF$_9$

Fig. 3. The comparison of BIC, function evaluations and model fitting time of sub-ARIMA with start parameters by different methods in dataset solar power.

Furthermore, from the in-depth analysis of the sub-ARIMA model with TSPs obtained from the different methods in Figs. 3 , 4, 5, 6 and 7, it has been observed that 1) models with TSPs have the same performance (BIC) but fewer function evaluations in the optimization and fitting time in model building than the Default in most cases, which implies that ARIMA models with TSPs from source domain perform better than self-built ARIMA model when no computation resources locally. 2) Models with TSPs have the same performance (BIC) but less fitting time in model fitting than those methods (i.e. HR, Innovations, and Kalman) compute on data itself in most cases though do not show fewer function evaluations in the optimization of all cases, which implies that ARIMA models with TSPs from source domain perform better than self-built ARIMA models that demand heavy computing and much time for start parameters. 3) TSPs improve the robustness of optimization of MLE of ARIMA because in those cases where other methods perform poorly, models with TSPs still have a good performance.

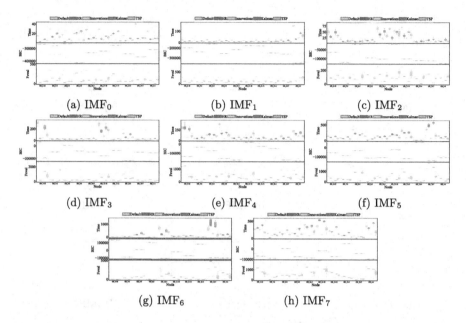

Fig. 4. The comparison of BIC, function evaluations and model fitting time of sub-ARIMA with start parameters by different methods in dataset Abilene.

To conclude, TSP from the domain model would help the target model fit with less time and still has the same performance, though they would not work in all cases because the optimization of MLE of ARIMA does not only rely on start parameters. The possible reason behind the failed ones is that 1) these sub-sequences have optimization problems such as saddle points and plateaus, 2) start parameters have minimal effect on optimization because no method has not been best in all cases either, 3) the common frequencies from the decomposition by the MVMD as similarity for transfer learning maybe not enough, which can be seen from the last two or three of IMFs of some datasets that have a relative high principle frequency (see the example of periodogram of dataset solar power in Fig. 2), which could be the cause of the poor performance of the MVMD in the separation of high frequencies [7,13]. Following the in-depth analysis of sub-sequences of these data owners, it has been observed that the better performance of MVMD-ARIMA with TSP belongs to those target time-series that have more prominent characteristics similar to domain time-series in the comparison of sub-sequences, which can be seen the IMFs that have more prominent common frequencies which mean more rigorous similarity.

3.4   Comparison of MVMD-Hybrid Framework

Based on the transfer effect, we can do collaboration modeling among different data owners in the same dataset, which means one data owner builds the MVMD-hybrid-TL model by building a sub-ARIMA model for only one IMF (or

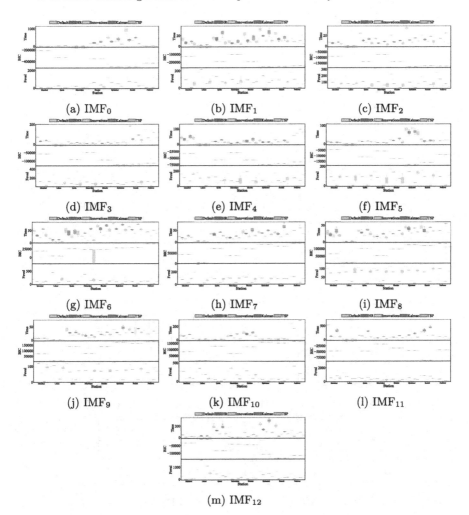

(a) IMF_0 (b) IMF_1 (c) IMF_2

(d) IMF_3 (e) IMF_4 (f) IMF_5

(g) IMF_6 (h) IMF_7 (i) IMF_8

(j) IMF_9 (k) IMF_{10} (l) IMF_{11}

(m) IMF_{12}

Fig. 5. The comparison of BIC, function evaluations and model fitting time of sub-ARIMA with start parameters by different methods in dataset GHI.

more IMFs in need) and then transfers between each other. Through this, we could get the MVMD-ARIMA-TL model with less fitting time and the same performance as self-built. To validate this, we use metrics mentioned above beside NRMSE, i.e. "time (sum)", "Feval (sum)" and "time (max)", and "Feval (max)" for successive and parallel MVMD-ARIMA-TL respectively. The comparison results are presented in Fig. 8a to Fig. 8e.

From Fig. 8a to Fig. 8e, it is observed that models with TSP have a close forecasting result (same NRMSE) as models with other methods but less fitting time. From the in-depth analysis, it can be seen that 1) MVMD-ARIMA models with TSP have a prominent performance in fitting time and function evalua-

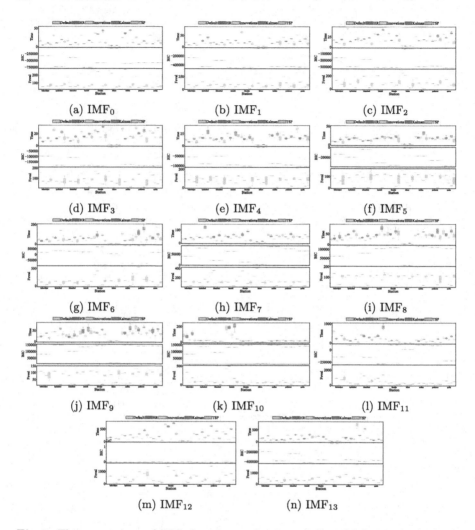

Fig. 6. The comparison of BIC, function evaluations and model fitting time of sub-ARIMA with start parameters by different methods in dataset wind speed.

tions than Default, which means collaboration among these similar time-series does help data owners that have no computation resources for start parameters, no matter using parallel (fewer max function evaluations and/or less max fitting time) or successively (fewer sum of function evaluations and/or less sum of fitting time) hybrid models. 2) MVMD-ARIMA models with TSP have a prominent performance in fitting time than those methods computed start parameters though not show fewer function evaluations in all cases, which means collaboration among these similar time-series does help data owners that have not enough computation resources or time for model building, no matter using parallel (fewer max function evaluations and/or less max fitting time) or successively

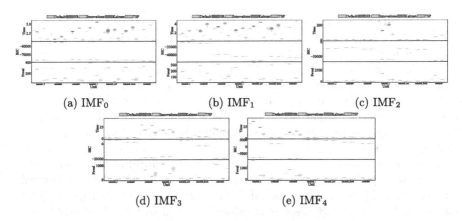

Fig. 7. The comparison of BIC, function evaluations and model fitting time of sub-ARIMA with start parameters by different methods in dataset PV.

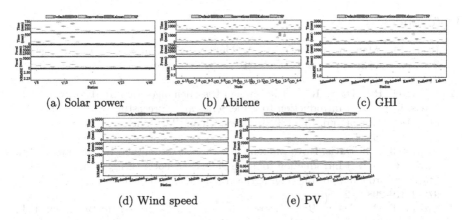

Fig. 8. The comparison of BIC, function evaluations and model fitting time of sub-ARIMA with start parameters by different methods in dataset PV.

(fewer sum of function evaluations and/or less sum of fitting time) hybrid models. The recorded improvement rises significantly higher in the case of these time series in one dataset that shows more same principle frequencies (see the example of periodogram of dataset solar power in Fig. 2). Hence, it can be concluded from the results that the proposed MVMD-hybrid-TL framework is reliable and effective at helping the MVMD-hybrid model with TSP in model fitting time or computing resources.

4   Conclusion

In this paper, we proposed a novel transfer learning framework (MVMD-hybrid-TL) that leverages the decomposition method MVMD to cluster sub-sequences of multiple time-series with similar characteristics and transfer knowledge across different domains. The framework can address the challenges of computing resource scarcity in decomposition-based hybrid models for multiple time-series forecasting. We conducted experiments on 5 real-world datasets from various domains such as energy consumption, network traffic, and solar radiation. The results showed that our framework achieved comparable performance with less fitting time than the baselines. Our work opens up new possibilities for applying transfer learning to decomposition-based hybrid models for multiple time-series prediction tasks.

There are still some works remained we will complement in future work such as the privacy preservation among the parameters transferred among multiple data owners and privacy-preserving MVMD to prevent privacy leakage in updating central frequencies. Besides, it is valuable as well to know how other kinds of models such as machine learning models or deep learning models would work in this transfer learning framework for decomposition-based hybrid models.

References

1. Bacher, P., Madsen, H., Nielsen, H.A.: Online short-term solar power forecasting. Sol. Energy **83**(10), 1772–1783 (2009). https://doi.org/10.1016/j.solener.2009.05.016
2. Brockwell, P.J., Davis, R.A.: The Hannan-Rissanen algorithm. In: Brockwell, P.J., Davis, R.A. (eds.) Introduction to Time Series and Forecasting. Springer Texts in Statistics, 3rd edn., pp. 137–139. Springer, Cham (2016). https://doi.org/10.1007/978-3-319-29854-2
3. Brockwell, P.J., Davis, R.A.: Maximum likelihood estimation. In: Brockwell, P.J., Davis, R.A. (eds.) Introduction to Time Series and Forecasting. Springer Texts in Statistics, 3rd edn., pp. 139–144. Springer, Cham (2016). https://doi.org/10.1007/978-3-319-29854-2
4. Brockwell, P.J., Davis, R.A.: Regression with ARMA errors. In: Brockwell, P.J., Davis, R.A. (eds.) Introduction to Time Series and Forecasting. Springer Texts in Statistics, 3rd edn., pp. 184–191. Springer, Cham (2016). https://doi.org/10.1007/978-3-319-29854-2
5. Byrd, R.H., Lu, P., Nocedal, J., Zhu, C.: A limited memory algorithm for bound constrained optimization. SIAM J. Sci. Comput. **16**(5), 1190–1208 (1995). https://doi.org/10.1137/0916069
6. Data, O.P.S.: Household data (2020). https://data.open-power-system-data.org/household_data/2020-04-15
7. Dragomiretskiy, K., Zosso, D.: Variational mode decomposition. IEEE Trans. Signal Process. **62**(3), 531–544 (2014). https://doi.org/10.1109/TSP.2013.2288675
8. Gonçalves, C., Bessa, R.J., Pinson, P.: A critical overview of privacy-preserving approaches for collaborative forecasting. Int. J. Forecast. **37**(1), 322–342 (2021). https://doi.org/10.1016/j.ijforecast.2020.06.003
9. Ju, C., et al.: Short-term hybrid forecasting model of wind power based on EEMD-ARIMA-LSTM. In: EMIE 2022; The 2nd International Conference on Electronic Materials and Information Engineering, Hangzhou, China, pp. 1–7 (2022)
10. Li, Z., Han, J., Song, Y.: On the forecasting of high-frequency financial time series based on ARIMA model improved by deep learning. J. Forecast. **39**(7), 1081–1097 (2020). https://doi.org/10.1002/for.2677

11. Lu, S., Zhang, Q., Chen, G., Seng, D.: A combined method for short-term traffic flow prediction based on recurrent neural network. Alex. Eng. J. **60**(1), 87–94 (2021). https://doi.org/10.1016/j.aej.2020.06.008

12. Powell, M.J.D.: An efficient method for finding the minimum of a function of several variables without calculating derivatives. Comput. J. **7**(2), 155–162 (1964). https://doi.org/10.1093/comjnl/7.2.155

13. Rehman, N.U., Aftab, H.: Multivariate variational mode decomposition. IEEE Trans. Signal Process. **67**(23), 6039–6052 (2019). https://doi.org/10.1109/TSP.2019.2951223

14. Sarmas, E., Dimitropoulos, N., Marinakis, V., Mylona, Z., Doukas, H.: Transfer learning strategies for solar power forecasting under data scarcity. Sci. Rep. **12**(1), 14643 (2022). https://doi.org/10.1038/s41598-022-18516-x

15. Tascikaraoglu, A., Uzunoglu, M.: A review of combined approaches for prediction of short-term wind speed and power. Renew. Sustain. Energy Rev. **34**, 243–254 (2014). https://doi.org/10.1016/j.rser.2014.03.033

16. World Bank: Pakistan - solar radiation measurement data (2017). https://energydata.info/dataset/pakistan-solar-radiation-measurement-data

17. Wright, M.: Direct search methods: once scorned, now respectable. In: Griffiths, D.F., Watson, G.A. (eds.) Numerical Analysis, pp. 191–208. Addison-Wesley, Harlow (1996)

18. Zhang, Y.: Abilene traffic matrices (2004). https://www.cs.utexas.edu/~yzhang/research/AbileneTM

Dataset Search over Integrated Metadata from China's Public Data Open Platforms

Qiaosheng Chen, Qing Shi, and Gong Cheng(✉) ⓘ

State Key Laboratory for Novel Software Technology, Nanjing University,
Nanjing, China
{qschen,qingshi}@smail.nju.edu.cn, gcheng@nju.edu.cn

Abstract. The aggregation, fusion, sharing, opening, development, and utilization of public data provide a solid basis for promoting the development of e-government and digital economy. These activities rely on an infrastructure software called public data open platform (PDOP) to provide enabling services. While China's national PDOP has yet to be completed, one pathway is to integrate the existing hundreds of provincial-level and prefectural-level PDOPs. However, these local PDOPs exhibit high heterogeneity, e.g., using different data catalogs and different metadata formats. In this system paper, we meet the challenge by crawling and integrating metadata records for datasets registered in existing local PDOPs, and we develop a prototype PDOP that provides unified search services over the integrated metadata. We conduct experiments to evaluate the core components of our prototype.

Keywords: Open government data · Public data open platform · Dataset search · Metadata integration

1 Introduction

Public data (公共数据) refers to the data in the public domain, typically including all the data collected or generated by government bodies. Public data is highly valuable and should be shared and reused, which is critical to the rapid development of digital economy. An established way of making public data available is to build a *public data open platform* (PDOP) which indexes the metadata of each open public dataset to be retrieved. Many countries in the world have developed their national PDOPs such as US government's data.gov and UK's data.gov.uk. In recent years, China has also promoted the construction of PDOPs. By September 2022, at least 21 provincial-level PDOPs have been built,[1] and there are also many prefectural-level PDOPs. The shared public data has already improved the effectiveness and efficiency of performing many and various data-driven tasks, e.g., in the prevention and control for COVID-19.

Motivation. To the best of our knowledge, China's national PDOP which has been designed to integrate all the local PDOPs has yet to be completed. Until

---

[1] https://www.gov.cn/zhengce/content/2022-10/28/content_5722322.htm.

E. Chen et al. (Eds.): BigData 2023, CCIS 2005, pp. 32–43, 2023.
https://doi.org/10.1007/978-981-99-8979-9_3

(a) Liaoning Province's PDOP

(b) Guangdong Province's PDOP

Fig. 1. Two provincial-level PDOPs providing different structures of data catalog for filtering search results and different metadata attributes in snippets.

the present, there is still a lack of an efficient way to search public data *across different local PDOPs in China*, which is challenging due to their *heterogeneity*. Indeed, existing provincial-level and prefectural-level PDOPs use their own data management systems. As illustrated in Fig. 1, they organize datasets into data catalogs of different structures, and their registered metadata records contain different attributes. It motivates us to meet the challenge and develop a prototype PDOP that integrates existing local PDOPs. Hopefully our solutions and lessons learned should be useful for completing China's national PDOP.

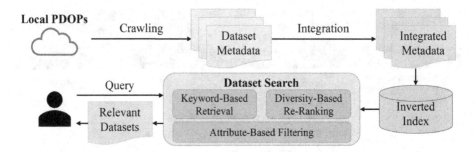

Fig. 2. Overview of our prototype.

Our Work. In this system paper, we present our ongoing work on building a PDOP that provides unified search services across existing local PDOPs in China. Figure 2 outlines its architecture. So far, we have crawled 562,038 metadata records for datasets registered in 124 local PDOPs, and have integrated heterogeneous metadata through rule-based attribute alignment and value cleaning, and BERT-based data catalog consolidation. Our current prototype provides keyword-based dataset retrieval with diversity-based re-ranking over the integrated and indexed metadata, and supports faceted search based on attribute filters. We have conducted experiments to evaluate the core components of our prototype and provide empirical insights for future research.

Prototype: http://open-data.store/cn-public

Structure of the Paper. In Sect. 2, we describe the crawling and integration of dataset metadata from existing local PDOPs. In Sect. 3, we introduce the implementation of dataset search services over integrated metadata. Experiments are presented in Sect. 4. Related work is discussed in Sect. 5, before we conclude the paper with future work in Sect. 6.

2 Crawling and Integration of Dataset Metadata

This section describes the methods employed for crawling dataset metadata from local PDOPs and integrating the crawled metadata in different formats.

2.1 Crawling of Dataset Metadata

At the time of writing the paper, we collected 124 provincial-level and prefectural-level PDOPs covering 25 provinces. Considering the different website structures of these PDOPs, we tailored our crawler for *each individual* PDOP. We crawled a total of 562,038 metadata records for datasets registered in these PDOPs.

Table 1. Most frequently used metadata attributes.

Attribute	Definition	Percentage
publisher (来源部门)	The entity responsible for making the dataset available	99.18%
title (标题)	A name given to the dataset	97.50%
access rights (开放类型)	Information about who can access the dataset	95.75%
description (描述)	A free-text account of the dataset	94.57%
update date (更新时间)	Most recent date on which the dataset was updated	93.56%
release date (发布时间)	Date of publication of the dataset	91.31%
format (数据格式)	The file formats of the dataset	90.95%
theme (所属主题)	A main category of the dataset	87.11%
record URL (详情页网址)	The URL of the record describing the registration of the dataset in a PDOP	84.48%
update frequency (更新频率)	The frequency at which the dataset is published	80.41%
size (数据规模)	The size of the dataset	61.17%
industry (所属行业)	The industry classification of the dataset	51.05%

2.2 Integration of Dataset Metadata

Attribute Alignment. Different PDOPs used heterogeneous metadata formats containing different attributes. To index and retrieve the crawled metadata records in a uniform way, we identified a set of core attributes that were commonly used, and then defined a set of *mapping rules* for each PDOP to convert its metadata format. Table 1 lists the most frequently used attributes. Seven attributes were notably used to describe more than 90% of the crawled datasets: publisher, title, access rights, description, update date, release date, and format.

Attribute Value Cleaning. We also cleaned the values of some attributes. For example, we converted access rights into a binary value: unconditionally or conditionally open. We converted all the dates into the YYYY-MM-DD format. We also consistently used filename extensions to represent file formats.

Data Catalog Consolidation. Considering that different PDOPs used heterogeneous data catalogs to organize datasets, to consolidate them into a single taxonomy to be used for filtering dataset search results, we used the 20 top-level categories of China's latest Industrial Classification for National Economic Activities (GB/T 4754-2017).[2] A dataset would be assigned a top-level category

---

[2] http://www.stats.gov.cn/sj/tjbz/gmjjhyfl/202302/P020230213400314380798.pdf.

if any of its sub-categories could match the theme or industry attribute of the dataset. String matching helped assign top-level categories to 300,111 datasets (53%) in this way. To assign top-level categories to each remaining dataset, we formulated it as a *multi-label classification* task and solved it by using a BERT-based classifier [12] fed with the title and description of the dataset. We trained the classifier on the assignment performed by string matching.

Fig. 3. Screenshot of a search results page returned by our prototype.

3   Dataset Search over Integrated Metadata

This section describes the design and implementation of our dataset search services over integrated metadata. Our prototype incorporates three search modules: keyword-based retrieval, diversity-based re-ranking, and attribute-based filtering. Figure 3 illustrates a screenshot of our search results page.

3.1   Keyword-Based Retrieval

Due to the lack of labeled data for training a powerful dense retriever, we implemented keyword-based retrieval by employing *field-weighted BM25* [21], a widely used unsupervised sparse retrieval model. Specifically, given a query q, for each dataset d, we computed the BM25 score of each of its fields $f_i \in F_d$, and then calculated a weighted sum over all the fields as the relevance score of d:

$$\text{rel}(d, q) = \sum_{f_i \in F_d} w_i \cdot \text{BM25}(f_i, q), \tag{1}$$

where w_i represents the weight of the i-th field f_i. We matched the query with seven selected metadata attributes as fields with weights in parentheses which

Fig. 4. Different metadata records for the same dataset collected from different PDOPs.

were tuned empirically: title (1.0), description (0.6), publisher (0.5), publisher's province (0.8), publisher's prefecture (0.8), theme (0.2), and industry (0.2).

Our implementation used Apache Lucene 9.6.0[3] for offline creating an inverted index based on which online retrieval would be fast. We used Lucene's SmartChineseAnalyzer for parsing and imported a list of stopwords from NLTK.[4]

As illustrated in Fig. 3, in search results pages, for each returned dataset we presented a *structured snippet* containing its selected metadata attributes where keyword matches were highlighted to help judge its relevance.

3.2   Diversity-Based Re-ranking

Different metadata records for the same dataset may be registered in multiple PDOPs, e.g., both in a provincial-level PDOP and in a prefectural-level PDOP as illustrated in Fig. 4, which requires deduplicating search results to improve diversify. Diversification would also help explore search results in a wider scope.

To achieve a trade-off between query relevance and diversity, inspired by the concept of *maximal marginal relevance* (MMR) [2], we re-ranked τ top-ranked search results, where τ was empirically set to 30 in our implementation. Specifically, given the τ most relevant datasets computed by Eq. (1), denoted by D_τ, we re-ranked them in an iterative manner. The dataset selected in the i-th iteration, denoted by d_i, was the dataset having the highest relevance score (i.e., rel) and the lowest similarity score (i.e., sim) to the datasets already selected in previous iterations:

$$d_i = \underset{d \in D_\tau \setminus \{d_1,\ldots,d_{i-1}\}}{\arg\max} \lambda \cdot \mathrm{rel}(d, q) - (1 - \lambda) \cdot \underset{d' \in \{d_1,\ldots,d_{i-1}\}}{\max} \mathrm{sim}(d, d'), \quad (2)$$

where the coefficient $\lambda \in [0, 1]$ was empirically set to 0.5 in our implementation.

We calculated an arithmetic mean of the similarities of five selected metadata attributes of two datasets as their similarity score. Specifically, for title and description, we computed *normalized edit distance* for such descriptive

---

[3] https://lucene.apache.org/.
[4] https://www.nltk.org/.

attributes. For example, for two dataset titles f_i, f_j, we calculated their similarity by

$$1 - \frac{\text{EditDistance}(f_i, f_j)}{\max\{|f_i|, |f_j|\}},$$

where $|\cdot|$ represents string length used to normalize edit distance into the range of $[0, 1]$. For publisher, publisher's province, and publisher's prefecture, we performed *strict string matching* for such nominal attributes. For example, for two publishers, their similarity would be either 1 representing string equality or 0 representing inequality.

3.3   Attribute-Based Filtering

As illustrated in Fig. 3, we implemented *faceted search* that allowed the user to filter the search results based on metadata attributes. Our current prototype supported four filters based on: publisher's province (省份), publisher's prefecture (城市), (consolidated) top-level industrial category (行业), and access rights (开放类型). The user could specify a value for each of these attributes to filter the search results.

4   Experiments

We conducted an empirical evaluation of the core components of our prototype.

4.1   Keyword-Based Retrieval

Retrieval Methods. We compared three popular sparse retrieval models: **BM25**, **TF-IDF** (based on cosine similarity), and **LMD** (short for Language Model using Dirichlet priors for smoothing). We used their implementations in Lucene.

Test Collection. Due to the lack of a standard test collection for evaluating dataset search over China's public data, we followed common practice [15, 18] to construct a test collection over our crawled dataset metadata. Specifically, we invited two domain experts to create 100 keyword queries about public data covering all the twenty top-level industrial categories, where 62 queries included the name of a province or prefecture. For each query, we used each of the three retrieval models to pool 20 top-ranked datasets. We pooled a total of 3,533 datasets. The domain experts manually annotated each pooled dataset as irrelevant (0), partially relevant (1), or relevant (2) as the gold standard. Unpooled datasets would be assumed to be irrelevant.

Evaluation Metrics. We used two standard metrics for evaluating retrieval accuracy: **NDCG@k** (short for normalized discounted cumulative gain) and **MAP@k** (short for mean average precision). When calculating MAP scores, partially relevant and relevant in the gold standard were both treated as relevant.

Table 2. Accuracy of different keyword-based retrieval methods.

	NDCG@5	NDCG@10	MAP@5	MAP@10
BM25	**0.7130**	**0.7070**	**0.3267**	**0.4369**
TF-IDF	0.6681	0.6763	0.2966	0.4152
LMD	0.5682	0.5797	0.2504	0.3474

Table 3. Diversity after re-ranking different numbers (τ) of search results.

	MeanSim@5	MeanSim@10	SimPairs@5	SimPairs@10	Time (ms)
BM25	0.4220	0.3529	3.74	12.16	8.72
+ MMR ($\tau = 10$)	0.2622	0.3529	1.23	12.16	88.20
+ MMR ($\tau = 20$)	0.1781	0.2301	0.38	4.54	323.34
+ MMR ($\tau = 30$)	0.1564	0.1936	0.30	2.72	716.14
+ MMR ($\tau = 40$)	0.1464	0.1781	0.26	2.02	1280.78
+ MMR ($\tau = 50$)	0.1374	0.1671	0.26	1.82	1986.89
TFIDF	0.4477	0.3889	4.09	14.21	44.38
+ MMR ($\tau = 10$)	0.3061	0.3889	1.76	14.21	122.22
+ MMR ($\tau = 20$)	0.2290	0.2815	0.94	7.30	349.02
+ MMR ($\tau = 30$)	0.2006	0.2400	0.65	5.13	747.40
+ MMR ($\tau = 40$)	0.1887	0.2199	0.57	4.13	1328.76
+ MMR ($\tau = 50$)	0.1827	0.2095	0.52	3.56	2045.92
LMD	0.3934	0.3320	3.59	12.47	43.27
+ MMR ($\tau = 10$)	0.2287	0.3320	1.36	12.47	120.75
+ MMR ($\tau = 20$)	0.1406	0.2081	0.39	4.87	348.28
+ MMR ($\tau = 30$)	0.1037	0.1513	0.08	1.63	777.90
+ MMR ($\tau = 40$)	0.0974	0.1336	0.07	0.83	1332.22
+ MMR ($\tau = 50$)	0.0917	0.1223	0.05	0.52	2051.02

Evaluation Results. Table 2 shows the mean scores of each method achieved over all the queries. The scores were basically at the same level as those observed on a comparable test collection [18]. BM25 outperformed the other two methods on all the metrics. TF-IDF achieved higher accuracy than LMD. Therefore, we chose BM25 to be used in our prototype.

4.2 Diversity-Based Re-ranking

Evaluation Metrics. We used two metrics to evaluate the diversity of search results after re-ranking different numbers (i.e., τ) of search results: **MeanSim@k** and **SimPairs@k**. The former measured the mean similarity (i.e., sim) between top-k search results. The latter represented the number of pairs of top-k search results whose similarity exceeded a threshold of 0.5. We also calculated the mean **time** used for responding to a query.

Evaluation Results. Table 3 shows the mean scores in each setting achieved over all the queries. By increasing τ to re-rank more search results, both MeanSim and SimPairs were reduced, i.e., diversity was improved, whereas the response time increased as expected. A reasonable trade-off between diversity and response time was observed at $\tau = 30$. Further increasing τ only marginally improved diversity but used more than one second to process a query. Therefore, we set $\tau = 30$ in our prototype.

4.3   Data Catalog Consolidation

Evaluation Setup. To evaluate the quality of data catalog consolidation, i.e., the accuracy of multi-label classification that assigned top-level industrial categories to a dataset based on its title and description, we randomly sampled 100 datasets and invited domain experts to manually check their predicted categories.

Evaluation Results. For 54% of the datasets, their predicted categories were judged to be totally correct. For 13% of the datasets, their predicted categories were partially correct, i.e., at least one but not all the predicted categories were correct. For 33% of the datasets, their predicted categories were incorrect. The results indicated room for further optimizing our classifier.

5   Related Work

5.1   National PDOPs in Other Countries

Most developed countries in the world have built their national PDOPs such as US government's data.gov, UK's data.gov.uk, Canada's open.canada.ca, and Australia's data.gov.au. Many of such PDOPs adopt a common data management system called CKAN,[5] based on which the metadata records registered in different PDOPs can be conveniently found, accessed, and interoperated in a consistent manner. For example, the European Open Data Portal [16] data.europa.eu has integrated metadata records collected from the national PDOPs of multiple European countries. Empirical studies have been conducted over these PDOPs. For instance, [14] analyzed the data requests in their query logs. and [17] characterized handcrafted summaries for datasets registered in data.gov.uk.

By contrast, China's national PDOP has yet to be completed, and existing local PDOPs seem to have adopted different data management systems, using heterogeneous data catalogs and metadata formats, making their integration a challenging task. Our work addresses this challenge with the design and implementation of a prototype PDOP that *provides a unified search service over metadata records crawled and integrated from different local PDOPs.*

---

[5] https://ckan.org/.

5.2   Dataset Search

Dataset search is a trending research topic, different from conventional keyword-based data search tasks [9,10,22], with researchers investigating various emerging research problems related to this topic [4]. A representative effort is Google's dataset search engine [1], which offers keyword-based retrieval over metadata records found on the web. In developing this system, a number of challenges have been identified, including the heterogeneity of metadata formats. Auctus [3] is a system that indexes the content summaries of tabular datasets and presents their data samples in search results pages. CKGSE [27,28], a prototype search engine for Chinese RDF datasets, offers content snippets [24,26] for each dataset to assist users in making informed relevance judgments.

Our prototype PDOP is *distinguished by the scope of data sources and the employed techniques*. Specifically, we fill the gap by crawling and integrating metadata records from a large (and increasing) number of China's local PDOPs with a unique focus on China's public data. From the technical perspective, we train a multi-label classifier to consolidate heterogeneous data catalogs, and perform diversity-based re-ranking to improve search results, both of which have not been addressed by existing PDOPs. We also report experimental results about the effectiveness and efficiency of the core components of our prototype.

6   Conclusion and Future Work

We have designed and implemented a prototype PDOP that represented the first step toward integrating China's local PDOPs. We crawled metadata records for China's public data from heterogeneous PDOPs, aligned and cleaned metadata attributes, consolidated data catalogs, and implemented unified keyword-based retrieval and faceted search services over integrated metadata. We carried out experiments to evaluate the core components of our prototype. Our solutions and experimental results are expected to be helpful for completing China's national PDOP which in turn would promote the development of China's digital economy.

We identified the following future directions. First, as shown by our experimental results, there is room for improving search accuracy. We plan to annotate more data to support training a more powerful dense retriever [6,13]. We will also explore better ways of consolidating data catalogs [23]. Second, observe that the quality of metadata is unsatisfying, e.g., for 33% of our crawled datasets, their descriptions trivially duplicate their titles. Data publishers are suggested to further improve metadata quality. Automated methods may also be developed to provide assistance, e.g., generating data summaries [8,11,20] and extracting representative data snippets [7,19,25]. Third, our current prototype is focused on crawling and processing dataset metadata, ignoring the actual data. Previous research has showed that incorporating dataset content could improve search accuracy and facilitate users' comprehension of datasets [5,18]. However, the large and heterogeneous data poses great challenges to be addressed.

Acknowledgements. This work was supported by the NSFC (62072224). The authors would like to thank all the students that helped crawl data.

References

1. Brickley, D., Burgess, M., Noy, N.F.: Google dataset search: building a search engine for datasets in an open web ecosystem. In: WWW 2019, pp. 1365–1375 (2019). https://doi.org/10.1145/3308558.3313685
2. Carbonell, J.G., Goldstein, J.: The use of MMR, diversity-based reranking for reordering documents and producing summaries. In: SIGIR 1998, pp. 335–336 (1998). https://doi.org/10.1145/290941.291025
3. Castelo, S., Rampin, R., Santos, A.S.R., Bessa, A., Chirigati, F., Freire, J.: Auctus: a dataset search engine for data discovery and augmentation. VLDB J. **14**(12), 2791–2794 (2021)
4. Chapman, A., et al.: Dataset search: a survey. VLDB J. **29**(1), 251–272 (2019). https://doi.org/10.1007/s00778-019-00564-x
5. Chen, J., Wang, X., Cheng, G., Kharlamov, E., Qu, Y.: Towards more usable dataset search: from query characterization to snippet generation. In: CIKM 2019, pp. 2445–2448 (2019). https://doi.org/10.1145/3357384.3358096
6. Chen, Q., et al.: Dense re-ranking with weak supervision for RDF dataset search. In: Payne, T.R., et al. (eds.) ISWC 2023, Part I. LNCS, vol. 14265, pp. 23–40. Springer, Cham (2023). https://doi.org/10.1007/978-3-031-47240-4_2
7. Cheng, G., Jin, C., Ding, W., Xu, D., Qu, Y.: Generating illustrative snippets for open data on the web. In: WSDM 2017, pp. 151–159 (2017). https://doi.org/10.1145/3018661.3018670
8. Cheng, G., Jin, C., Qu, Y.: HIEDS: a generic and efficient approach to hierarchical dataset summarization. In: IJCAI 2016, pp. 3705–3711 (2016)
9. Cheng, G., Li, S., Zhang, K., Li, C.: Generating compact and relaxable answers to keyword queries over knowledge graphs. In: Pan, J.Z., et al. (eds.) ISWC 2020, Part I. LNCS, vol. 12506, pp. 110–127. Springer, Cham (2020). https://doi.org/10.1007/978-3-030-62419-4_7
10. Cheng, G., Qu, Y.: Searching linked objects with falcons: approach, implementation and evaluation. Int. J. Semantic Web Inf. Syst. **5**(3), 49–70 (2009). https://doi.org/10.4018/jswis.2009081903
11. Cheng, G., Tran, T., Qu, Y.: RELIN: relatedness and informativeness-based centrality for entity summarization. In: Aroyo, L., et al. (eds.) ISWC 2011, Part I. LNCS, vol. 7031, pp. 114–129. Springer, Heidelberg (2011). https://doi.org/10.1007/978-3-642-25073-6_8
12. Devlin, J., Chang, M., Lee, K., Toutanova, K.: BERT: pre-training of deep bidirectional transformers for language understanding. In: NAACL-HLT 2019, vol. 1. pp. 4171–4186 (2019). https://doi.org/10.18653/v1/n19-1423
13. Guo, J., Cai, Y., Fan, Y., Sun, F., Zhang, R., Cheng, X.: Semantic models for the first-stage retrieval: a comprehensive review. ACM Trans. Inf. Syst. **40**(4), 66:1–66:42 (2022). https://doi.org/10.1145/3486250
14. Kacprzak, E., Koesten, L., Ibáñez, L., Blount, T., Tennison, J., Simperl, E.: Characterising dataset search - an analysis of search logs and data requests. J. Web Semant. **55**, 37–55 (2019). https://doi.org/10.1016/j.websem.2018.11.003
15. Kato, M.P., Ohshima, H., Liu, Y., Chen, H.O.: A test collection for ad-hoc dataset retrieval. In: SIGIR 2021, pp. 2450–2456 (2021). https://doi.org/10.1145/3404835.3463261

16. Kirstein, F., Dutkowski, S., Dittwald, B., Hauswirth, M.: The European data portal: scalable harvesting and management of linked open data. In: ISWC 2019 Satellite Tracks, pp. 321–322 (2019)

17. Koesten, L., Simperl, E., Blount, T., Kacprzak, E., Tennison, J.: Everything you always wanted to know about a dataset: studies in data summarisation. Int. J. Hum. Comput. Stud. **135** (2020). https://doi.org/10.1016/j.ijhcs.2019.10.004

18. Lin, T., et al.: ACORDAR: a test collection for ad hoc content-based (RDF) dataset retrieval. In: SIGIR 2022, pp. 2981–2991 (2022). https://doi.org/10.1145/3477495.3531729

19. Liu, D., Cheng, G., Liu, Q., Qu, Y.: Fast and practical snippet generation for RDF datasets. ACM Trans. Web **13**(4), 19:1–19:38 (2019). https://doi.org/10.1145/3365575

20. Liu, Q., Cheng, G., Gunaratna, K., Qu, Y.: Entity summarization: state of the art and future challenges. J. Web Semant. **69**, 100647 (2021). https://doi.org/10.1016/j.websem.2021.100647

21. Robertson, S., Zaragoza, H.: The probabilistic relevance framework: BM25 and beyond. Found. Trends Inf. Retr. **3**(4), 333–389 (2009). https://doi.org/10.1561/1500000019

22. Shi, Y., Cheng, G., Kharlamov, E.: Keyword search over knowledge graphs via static and dynamic hub labelings. In: WWW 2020, pp. 235–245 (2020). https://doi.org/10.1145/3366423.3380110

23. Shvaiko, P., Euzenat, J.: Ontology matching: state of the art and future challenges. IEEE Trans. Knowl. Data Eng. **25**(1), 158–176 (2013). https://doi.org/10.1109/TKDE.2011.253

24. Wang, X., Cheng, G., Kharlamov, E.: Towards multi-facet snippets for dataset search. In: PROFILES & SEMEX 2019, pp. 1–6 (2019)

25. Wang, X., et al.: PCSG: pattern-coverage snippet generation for RDF datasets. In: Hotho, A., et al. (eds.) ISWC 2021. LNCS, vol. 12922, pp. 3–20. Springer, Cham (2021). https://doi.org/10.1007/978-3-030-88361-4_1

26. Wang, X., Cheng, G., Pan, J.Z., Kharlamov, E., Qu, Y.: BANDAR: benchmarking snippet generation algorithms for (RDF) dataset search. IEEE Trans. Knowl. Data Eng. **35**(2), 1227–1241 (2023). https://doi.org/10.1109/TKDE.2021.3095309

27. Wang, X., Lin, T., Luo, W., Cheng, G., Qu, Y.: Content-based open knowledge graph search: a preliminary study with OpenKG.CN. In: Qin, B., Jin, Z., Wang, H., Pan, J., Liu, Y., An, B. (eds.) CCKS 2021. CCIS, vol. 1466, pp. 104–115. Springer, Singapore (2021). https://doi.org/10.1007/978-981-16-6471-7_8

28. Wang, X., Lin, T., Luo, W., Cheng, G., Qu, Y.: CKGSE: a prototype search engine for Chinese knowledge graphs. Data Intell. **4**(1), 41–65 (2022). https://doi.org/10.1162/dint_a_00118

Integrating DCNNs with Genetic Algorithm for Diabetic Retinopathy Classification

Zhengfu Li[1], Liping Wu[1](✉), and Jiaojiao Li[2]

[1] School of Computer Engineering, Jiangsu Ocean University,
Lianyungang 222005, China
1677610760@qq.com
[2] School of Pharmacy, Jiangsu Ocean University, Lianyungang 222005, China

Abstract. Diabetic retinopathy (DR) is a major ocular complication of diabetes. Delayed diagnosis of such disease increases the risk of vision loss and irreversible blindness. With the popularization of computer-aided diagnosis technology, the use of deep learning for DR classification has become a current research hotspot. We aim to develop a GA-DCNN model that can improve the performance of DR classification. In this paper, a novel global attention-based model called GCA-SA is proposed to provide fine-grained global lesion information for DR classification. Furthermore, inspired by genetic algorithm (GA) and ensemble learning (EL), this paper also proposes a strategy of integrating deep convolutional neural networks (DCNNs) with GA. The GA-DCNN model is constructed by aggregating GCA-SA and spatial pyramid pooling (SPP) into three DCNNs and using the strategy of integrating DCNNs with GA. The experimental results show that the accuracy, specificity and AUC of the GA-DCNN reach 0.91, 0.94 and 0.93, respectively. Compared with traditional CNN, GA-DCNN can capture the detailed features of DR lesions and integrate the classification results of the multiple DCNNs, effectively improving the detection and classification performance of DR.

Keywords: Diabetic Retinopathy · DCNN · Global Attention · Genetic Algorithm · Ensemble Learning

1 Introduction

Diabetic Retinopathy (DR) is a severe chronic eye disease whose incidence is closely correlated with the dramatic rise in the global number of diabetes patients [1]. According to statistics, there are about 460 million people with diabetes in the world, and it is expected to reach 700 million by 2045 [2,3]. Given these issues and limitations, this study aims to develop a more efficient and accurate DR classification model. The retinal image is shown in Fig. 1.

At present, the pathogenesis of DR is not very clear, but the early diagnosis, drug development and treatment of this retinal disease have attracted the

E. Chen et al. (Eds.): BigData 2023, CCIS 2005, pp. 44–60, 2023.
https://doi.org/10.1007/978-981-99-8979-9_4

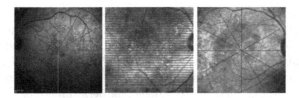

Fig. 1. The example of retinal image.

attention of the medical and industrial circles [4]. In recent years, convolutional neural networks (CNNs) have been continuously developed and have achieved impressive achievements in various tasks of image processing, classification and detection [5]. In addition, medical image classification based on deep learning has entered the stage of clinical experiments, accelerating the development of computer-aided diagnosis systems [6]. For DR classification, Al-Antary et al. proposed a multi-scale CNN model that extracts feature maps of fundus images at different scales for DR classification [7]. Krause et al. used the InceptonV4 network to train many fundus images and improved the performance of DR classification [8]. Qomariah et al. proposed a CNN with transfer learning and used a support vector machine algorithm for DR classification, which improved the accuracy of CNN classification [9]. Yang et al. proposes a two-stage approach that utilizes a lesion detection module and a lesion classification module to achieve a bidirectional exchange of lesion information and image-level information and implements the fine-grained classification task of DR [10].

While deep learning, specifically Convolutional Neural Networks (CNNs), has made significant strides in the image classification of DR, it still faces various challenges in clinical applications. On the one hand, the problem of missing labels on datasets leads to an imbalance of positive and negative samples, and the normal physiological features in fundus images also interfere with the diagnosis of DR lesions. On the other hand, retinal diseases have visual similarity. In this regard, an efficient attention module is used to capture the specific lesion features of DR. Kamran et al. proposed a method of connecting two sub-networks [11]. One network is used for disease encoding in a supervised manner, and the other network is used for disease attention graph generation in an unsupervised manner, which improves the robustness of CNN. Ding et al. proposed an attention pyramid network that can learn high-level abstract features and low-level detailed features to achieve accurate localization of lesion regions [12]. Existing DR classification models often overlook the fine-grained features of the disease, limiting their efficacy in realworld medical settings. To further improve the performance of DR classification, this paper proposes a GA-DCNN model that introduces the novel GCA-SA to highlight key DR features. The classification results are integrated by genetic algorithm (GA) with deep convolutional neural networks (DCNNs). Furthermore, transfer learning strategies are used to transfer general information to the DR classification domain, improving the generalization ability of DCNN. The main contributions of this paper can be summarized as follows:

- This paper proposes a GCA-SA that aggregates GCA-SA and spatial pyramid pooling (SPP) [13] into DenseNet201, MobileNet, and InceptionV3. GCA-SA is used to capture more identifiable DR lesion information, and SPP is used to receive features of different scales, making up for the defect that traditional networks only receive fixed-dimensional data. Therefore, the combination of the two structures can improve the detection performance of DR.
- Inspired by the ensemble learning and the global optimization characteristics of GA, this paper also proposes a strategy of integrating DCNNs with GA and synthesizes the classification results of three DCNNs to improve the performance of DR classification.
- Based on the strategy of integrating DCNNs with GA, GA-DCNN is constructed. The experimental results on the DR dataset show that the accuracy, specificity, and AUC reach 0.91, 0.94, and 0.93, respectively. GA-DCNN can not only improve the detection performance of DR, but also the classification performance of DR.

The rest of this paper is organized as follows. Section 2 reviews the related work on deep learning algorithms for DR classification. Section 3 elaborates on the GA-DCNN model in detail. In Sect. 4, the classification performance of GA-DCNN is verified by experiments and compared with other classical algorithms. Sections 5 provides conclusions.

2   Related Work

As an end-to-end feature extractor, CNN can automatically extract subtle lesion features in fundus images to complete the DR classification task. In this section, the recent research on DR classification based on deep learning methods is briefly reviewed.

2.1   Single CNN for DR Classification

There are two main types of single CNN used for DR classification. The first is based on lesion ROI annotation. Xia et al. proposed a multi-scale segmentation and classification model based on MSRNet and MS-EfficientNet to achieve accurate detection and classification of microaneurysms [14]. Ramya et al. proposed a hybrid convolutional neural network combining binary local search optimizer and particle swarm optimization algorithm for hyperparameter optimization and then achieved excellent DR classification performance on the public ROC dataset and ARA400 dataset [15]. Eftekheri et al. [16] proposed a two-stage CNN framework for DR detection and classification. First, a detector is used to select candidate regions of MA from the ROI. Then, a classifier is used to separate MA and non-MA, which improves the detection performance of MA. The second is based on image-level annotation. Gulshan et al. used InceptionV3 for DR classification and experimented with the EyePACS dataset with high specificity and sensitivity as evaluation indicators [17]. The sensitivity and specificity of InceptionV3 for detecting diabetic retinopathy outperform existing models. Lin et al. designed a multi-task framework based on the attention mechanism module [18], which first connected two subtasks, i.e., detection and classification, and then separated the two subtasks, which enhanced the performance of DR classification. He et al. proposed two novel attention mechanism modules [19]. The former explores more discriminative regional features, while the latter captures subtle lesion information. The two modules were aggregated into the backbone network to form CABNet for DR classification. Experimental results show that CABNet significantly improves the performance of DR classification. Previous work has shown that deep learning is effective for DR classification. The task of DR classification remains challenging given the limitations of a single CNN decision.

2.2 Multiple CNNs for DR Classification

To further improve the classification performance of deep learning algorithms, some researchers try to change the classification structure of a single model in the past, and use multiple models for parallel training. DR classification algorithms based on multiple CNNs can be roughly divided into two categories. The first is DR classification through multi-model feature fusion. Nneji et al. proposed a weighted fusion-based deep learning model (WFDLN), which was trained on InceptionV3 and VGG16 using CLAHE and CECED processed fundus images, respectively [20], and weighted fusion of the output feature maps of the two channels. Experiments on the Messidor dataset and Kaggle dataset show that WFDLN can achieve high precision for DR classification. Li et al. proposed a lesion-attention pyramid network (LAPN), in which images of three different resolutions were used as the input of the pyramid sub-network, and the lesion attention module (LAM) was used to fuse the features of high resolution, low resolution and lesion activation maps [21]. The experimental results show that this method is superior to other existing methods, and the lesion activation map with lesion consistency can be used as supplementary evidence for the clinical diagnosis of DR. Zhao et al. proposed a novel deep fusion network, which fused features of two CNNs to achieve higher accuracy than single CNN [22]. At the same time, there are fewer network layers with the same number of parameters. The second approach is to synthesize the decisions of multiple sub-categories through ensemble learning. Gao et al. ensembled models based on multiple CNNs [23]. The method first generates three subsets from the Kaggle dataset and then designed two ensemble learning schemes. In the first scheme, DensNet121 is used to train three subsets, and the classification results were then integrated. The second scheme is to train three subsets using ResNet50, DensNet121 and InceptionV3, respectively, and then integrate the classification results. The experimental results show that the ensemble learning strategy can improve classification performance. Zhuang et al. proposed a DR detection algorithm (DR-IIXRN) based on deep ensemble learning and attention mechanism [24]. In this model, the severity of DR was initially determined by improved Inception V3, InceptionResNet V2, Exception, ResNeXt101, and NASNetLarge, and then a weighted voting algorithm was used to achieve the five-class classification of DR. The proposed GA-DCNN uses image-level labels without additional lesion location information. Most importantly, a global attention module is introduced to suppress irrelevant feature information, and the strategy of DCNNs integrated with GA is used to synthesize the classification results, which improves the classification performance of the model.

3   Methodology

The structure of the proposed GA-DCNN is shown in Fig. 2. The GA-DCNN model mainly consists of two parts. The first part uses three DCNNs to extract

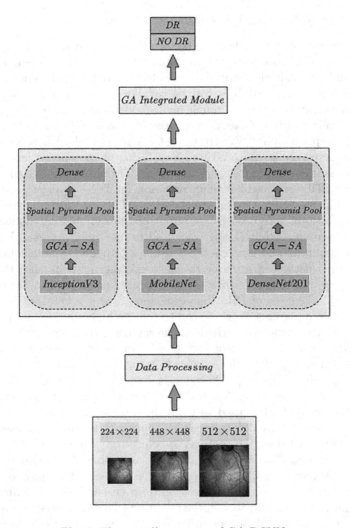

Fig. 2. The overall structure of GA-DCNN.

the lesion information in the fundus images, and then uses the GCA-SA module to obtain the specific expression of the DR feature, and uses the classifier to generate the prediction result. The second part adds a GA module based on the first part. The function of the module is to make comprehensive decisions on the classification results of the three DCNNs through the strategy of integrating DCNNs with GA, thereby improving the performance of DR detection and classification.

3.1 Overview of GA-DCNN

As shown in Fig. 2, GA-DCNN uses the three DCNNs (DenseNet201, MobileNet and InceptionV3) after migration learning on the ImageNet dataset as the backbone of feature extraction. It can be seen from Fig. 1 that the feature information related to the DR classification task only occupies a small part of the entire retinal image (including microaneurysm, hemorrhage, exudate, etc.), or even several small lesion points. The number of convolutional layers and the size and number of convolutional kernels of DCNN have great influence on the extraction of lesion features. Images with single resolution may lose lesion features due to frequent convolution operations. This paper takes images with three resolutions (224×224, 448×448 and 512×512) as input for image preprocessing, making the DCNN can not only learn high-resolution global feature information, but also learn low-resolution detailed feature information. However, fundus images have visual similarity and easily confused with the normal physiological structure. To enhance the attention to key lesion features, the multi-scale lesion features extracted from the network are put into the GCA-SA module, enabling the DCNN to adaptively focus on the most recognizable regions in the feature map. At the moment, the dimensions of the output features of the model are inconsistent, and it is difficult to directly put into the full connection layer for classification operations. SPP can calculate a single feature map from the image, and fuses local features in any subregion of the image to generate vectors of fixed length, so as to avoid repeated calculation. The spatial pyramid pooling structure is introduced before the fully connected layer, so that DCNN can accept multi-scale images and get rid of the limitation of the vector dimension by the fully connected layer. Since GA has great global optimization properties, it can obtain optimal results through multiple iterations. Therefore, we introduce GA into the ensemble learning model and propose a strategy to integrate GA with DCNNs to synthesize the classification results. GA-DCNN firstly combines DCNN with GCA-SA and then integrates three improved DCNNs with GA. The former can improve the ability of model feature extraction, while the latter can improve the decision-making performance of the model, thereby promoting the improvement of DR classification performance.

3.2 GCA-SA Module

The core idea of the GCA-SA module is based on the fundus feature map extracted by DCNN to obtain a more distinguishable feature expression of the

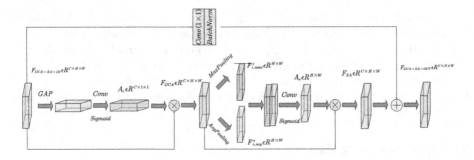

Fig. 3. Overview of the GCA-SA module.

lesions. The feature maps extracted by DCNN only include high-level expression of fundus images, while DR classification needs to capture the specific expression of lesion features. Feature maps of different scales have different feature expression abilities, and simple feature stacking cannot take advantage of feature expression. The attention mechanism weights the original feature map by spatial weight, which can effectively make up for the defect of DCNN's feature expression ability. The designed GCA-SA module mainly includes two parts. First, the feature representation of each channel is obtained according to the channel attention mechanism, and the attention map of the channel direction is generated. Second, the spatial feature representations of the lesions are found according to the feature maps captured by the channel attention module, and then the spatial orientation attention maps are generated.

Figure 3 illustrates the detailed structure of the GCA-SA module, which adopts both the channel-attention mechanism and the spatial-attention mechanism to generate lesion-related attention feature maps. Different from the CBAM module in [25], the channel-attention module adopts a single-branch structure, while the spatial-attention module adopts a double-branch structure. First, the channel-attention weights are calculated by the following formula:

$$A_c = \sigma(Conv(GAP(F_{GCA-SA-IN})))\tag{1}$$

where σ is the sigmoid activation function, GAP is the global average pooling, and $Conv$ is the convolutional operation. After obtaining the weight of the channel-attention A_c, it is multiplied by the input feature map to get the lesion feature in the channel direction:

$$F^i_{GCA} = A_c \otimes F_{GCA-SA-IN} \tag{2}$$

where $\otimes$ denotes the product of the corresponding elements. Therefore, the channel-attention module can acquire DR-related channel features and suppress DR-unrelated redundant features in the channel direction. The GCA-SA module aggregates two spatial feature representations using average pooling and max pooling operations along the spatial direction, and then calculates the spatial-attention weight A_s :

$$A_s = \sigma(Conv([F^s_{i,\max}; F^s_{i,avg}])) \tag{3}$$

Then, it is multiplied with the output of the channel attention module to generate lesion features in the spatial orientation:

$$F^i_{SA} = A_s \otimes F^i_{GCA} \tag{4}$$

The spatial-attention mechanism can acquire DR-related spatial features and suppress DR-unrelated redundant features in the spatial direction. Finally, the input and output features of the GCA-SA module are aggregated through residual connections, and the output feature map is calculated as follows:

$$F_{GCA-SA-OUT} = BN(Conv(F_{GCA-SA-IN})) \oplus F_{SA} \tag{5}$$

where $\oplus$ denotes the addition of the corresponding elements. Therefore, the GCA-SA module captures the correlation between the DR lesions of the channel and the spatial orientation by updating the weight information.

3.3   The Strategy of Integrating DCNNs with GA

Genetic algorithm is a heuristic global search and optimization algorithm that simulates biological evolution mechanism [26]. Ensemble learning (EL) can synthesize the decision results of multiple models according to certain rules [27]. The typical approach for integrating ensemble learning involves taking the average of the output probabilities from multiple base classifiers to yield the prediction probabilities of the final model. Due to the inconsistent classification capabilities of each base classifier, the simple averaging method cannot directly extract the classification capabilities of each base classifier. This study proposes a strategy for integrating DCNNs with GA, and the process is shown in Table 1. Considering the value of λ in Table 1 used to balance the performance of the models, this paper uses the genetic algorithm to iterate it several times to obtain the optimal parameter settings. The process is shown in Table 2.

Table 1. Algorithm 1

Algorithm 1: The strategy of integrating DCNNs with GA

Input: Training dataset D, Testing dataset T, models DCNNs = [M1, M2, M3]
Output: Prediction results P
1: **Step 1:** Calculate the AUC of each DCNN
2: AUC_values = calculate_auc(D, DCNNs)

3: **Step 2:** Calculate the weight of each DCNN
4: Weight_values = calculate_weight(AUC_values,λ)

5: **Step 3:** Normalize the weight value
6: Weight_normalize = normalize(Weight_values)

7: **Step 4:** Obtain the prediction probability value
8: Test_probability = calculate_test_probability(T, DCNNs)

9: **Step 5:** Calculate the final predicted probability value
10: Test_predict_probability = calculate_classification_probabality
(Test_probability, Weight_normalize)

11: **Step 6:** Calculate the prediction category
12: Test_predict_results = calculate_prediction_results(Test_predict_probability)

For Table 1, after data training, calculate the AUC values of each DCNN and put them into AUC_values. Then, calculate the weights by introducing λ and normalize them. Finally, apply the model to the testset to obtain prediction probabilities, and combine them with the weight values to optimize the prediction probabilities, thereby achieving the purpose of correcting prediction results. The formula for calculating the weight of each DCNN is as follows:

$$\omega_i = \frac{AUC\_values[i]}{(\sum_{j=1}^{k} AUC\_values[j])} * \lambda_i \tag{6}$$

where, k is the number of DCNN models, λ_i is a parameter used to balance the classification performance of the model, $\lambda_i \in [1,10]$. In addition, the predicted probability value of each DCNN is weighted and summed to calculate the final predicted probability value. The calculation formula is as follows:

$$y = \sum_{i=1}^{k} \omega_i * Test\_probability[i] \tag{7}$$

where, $Test\_probability$ is the predicted probability of testset. For Table 2, this paper uses GA to iterate and optimize in Table 1 and to further enhance the classification performance of the model. First, the accuracy rate is taken as the fitness function, and the roulette and elite retention strategies are combined to perform the selection operation. The calculation formula is as follows:

Table 2. Algorithm 2

Algorithm 2: Genetic Algorithm

Input: Testing dataset T, models DCNNs = [M1, M2, M3],
initialize GA parameters GA_param
Output: λ value for each testset sample $\lambda = [\lambda_1, \lambda_2, \lambda_3]$
1: for gen in total_genes:
2:   **Step 1:** Initialize the population for GA
3:   population = create_new_population(GA_param)
4:   **Step 2:** Evaluate fitness of each individual in the population
5:   fitness = calculate_fitness(population, T, DCNNs)
6:   **Step 3:** Select individuals for reproduction (mating) based on their fitness
7:   parents = select_population(population, fitness)
8:   **Step 4:** Generate new population through crossover and mutation
9:   population = crossover_and_mutation(parents)
10: **Step 5:** Update parameters of λ
11: λ = update_params(population)
12: **Step 6:** Judge whether or not it meets the termination condition of iteration
13: if stop_criteria(fitness):
14:      break
15: Return λ

$$p_i = \frac{f_i}{\sum\limits_{k=1}^{N} f_k} \tag{8}$$

where f_i is the i th fitness value, and p_i is the i th selection probability. Then, according to the crossover strategy, two parental individuals are randomly selected from the current generation, and two new chromosomes were generated by single-point crossover. Suppose the individual genes chosen at random are v_1 and v_2. The calculation formula is as follows:

$$\begin{cases} c_1 = \alpha v_2 + (1-\alpha)v_1 \\ c_2 = \alpha v_1 + (1-\alpha)v_2 \end{cases} \tag{9}$$

where α is random parameters, c_1 and c_2 are genes of newly generated individuals. Finally, according to the mutation strategy, an individual is randomly selected from the current generation, and then the gene on the individual is updated. The calculation formula is as follows:

$$\begin{cases} c = v + (1-v) * (1 - r^{(1-i/iter_{\max})^2}) & r > 0.5 \\ c = v - (v-0) * (1 - r^{(1-i/iter_{\max})^2}) & r \leq 0.5 \end{cases} \tag{10}$$

4   Experiment Results

4.1   Dataset

The DR dataset in this paper comes from the 2021 Asia-Pacific Society of Ophthalmology Big Data Competition, and the Alibaba Cloud Tianchi platform provides exclusive technical support for the competition [28]. In order to meet the experimental requirements of DR classification, 1200 fundus images before treatment were screened from the original dataset and divided into the training set and test set according to the ratio of 8:2. Before the experiment, we need to crop, scale, standardize and separate color channels on the original image. In order to meet the requirements of multi-scale feature extraction, the resolutions of the images are adjusted to 224×224, 448×448 and 512×512, respectively.

4.2   Evaluation Metrics

In this paper, images in the dataset are labeled as 0 (no DR) and 1 (with DR), and the performance of the model is evaluated in terms of Accuracy, Specificity and Sensitivity. AUC is a performance indicator for classification problems under different threshold settings.

4.3   Results and Analysis

To evaluate the DR classification performance of the GA-DCNN model, this paper divides the experiments into three groups. The first group is to verify the effect of different resolution fundus images on DCNN training, and the second group is to verify the effect of the GCA-SA module on DCNN classification performance. The third set of experiments is to verify the effect of the strategy of integrating DCNNs with GA for DR classification.

Due to the introduction of the spatial pyramid pooling structure in DCNN, different sizes of pooling kernels are tested. With other parameters fixed, different pooling kernels are introduced into three different DCNNs, and the average accuracy of multiple experiments is taken as the final result.

Table 3. Comparison of the effect of the kernel size of SPP on model performance

Methods	Kernel size of SPP	Accuracy (%)	Sensitivity (%)	Specificity(%)
Dense-Net201	[1,2,5]	84.8	85.6	83.7
	[1,2,3]	85.8	86.2	85.3
	[1,2,4]	**86.0**	**86.7**	**85.8**
MobileNet	[1,2,3]	80.5	80.9	80.1
	[1,2,5]	85.2	85.8	84.7
	[1,2,4]	**87.2**	**87.6**	**86.8**
InceptionV3	[1,2,5]	85.2	85.5	84.6
	[1,2,4]	85.8	86.0	85.1
	[1,2,3]	**86.3**	**86.6**	**85.5**

Table 3 shows the influence of pooling kernel size on model performance. DenseNet201 and MobileNet work better when the pooling kernel parameter is set to [1,2,4], while InceptionV3 works better when the pooling kernel parameter is set to [1,2,3]. It can be seen that different pooling kernels have certain differences in the classification effects for different CNNs. Therefore, setting appropriate pooling kernel parameters is very important for the design of DCNN models.

Experiment 1: To verify the effectiveness of neural network model training on different resolution images, experiments were conducted on DenseNet201, MobileNet and InceptionV3. There are inter-class similarities in fundus images, while subtle differences exist within classes. In other words, abnormal images and normal images are visually indistinguishable, but there are significant differences. Usually, image classification needs to compress image data to a fixed size, but the original resolution of medical images is large, and excessive compression will lead to information loss when DCNN extracts features.

Therefore, in Experiment 1, images with different resolutions are used as input, and DCNN is used to extract multi-scale features to compensate for the loss of information. To prove the above speculation, the following comparative experiment was designed. First, we divided the experimental data into two groups: the first group was single-resolution images (fundus images of the same resolution), and the second group was multi-resolution images (fundus images of three different resolutions). Then, single-resolution images and multi-resolution images are used as inputs to the three DCNNs, respectively. The pooling kernel size of the SPP structure is set as [1,2,4]. The results are shown in Table 4.

Table 4. Comparison of model classification results with single-resolution and multi-resolution images as inputs

Resolution	Methods	Accuracy (%)	Sensitivity(%)	Specificity(%)
Single-resolution	InceptionV3	79.3	79.8	78.5
	DenseNet201	84.3	84.7	83.8
	MobileNet	84.5	84.9	84.1
Multi-resolution	InceptionV3	85.8	86.0	85.1
	DenseNet201	86.0	86.7	85.8
	MobileNet	87.2	87.6	86.8

Table 4 shows the classification performance of DR for images of different resolutions. It can be seen that our speculation is correct. Compared to single-resolution images, the performance of the DCNN model is significantly improved when multi-resolution images are used as input. At the same time, the training results of DenseNet201 are significantly improved by multi-resolution images, while the training performance of MobileNet and InceptionV3 are greatly improved. The experimental results prove that, due to the small lesion area of medical images, DR classification emphasizes fine-grained feature information, and multi-resolution images are beneficial to DCNN to learn features of different scales and achieve accurate detection of lesion information.

Experiment 2: To verify the influence of the GCA-SA module on the DR classification performance, the GCA-SA module was introduced into the original DCNN based on Experiment 2, and the pooling kernel parameters of SPP were fixed as [1,2,4]. It can be seen from Table 5 that under the same experimental environment, the GCA-SA module can achieve good performance in the DR classification task. The experimental results show that without the introduction of the GCA-SA module, the accuracy of the three DCNNs increases from 84.3%, 84.5% and 79.3% to 85.0%, 85.3% and 80.5%, respectively. After introducing the GCA-SA module, the accuracies of the three DCNNs improved from 86.0%, 87.0% and 85.8% to 87.5%, 88.0% and 87.9%, respectively. In addition, in order to further verify the effect of GCA-SA, CBAM and GCA-SA are respectively added to different DCNN models. The results show that GCA-SA has better performance. Therefore, the GCA-SA module can automatically mine more identifiable lesion features in medical images, remove redundant features unrelated to DR and improve DR detection performance.

Table 5. Comparison of the influences of the GCA-SA module on DCNN performance

Methods	SPP	CBAM	GCA-SA	Accuracy (%)	Sensitivity(%)	Specificity(%)
DenseNet201				84.3	84.7	83.9
	✓			84.5	84.8	84.2
		✓		85.0	85.5	84.3
	✓			86.0	86.3	85.6
	✓	✓		86.3	86.7	86.0
	✓		✓	**87.5**	**87.9**	**87.2**
MobileNet				84.5	84.7	84.1
	✓			84.9	85.3	84.2
		✓		85.3	85.9	84.7
	✓			87.0	87.6	86.4
	✓	✓		87.2	87.8	86.7
	✓		✓	**88.0**	**88.6**	**87.3**
InceptionV3				79.3	79.8	79.1
	✓			80.0	80.4	79.7
		✓		80.5	80.9	80.2
	✓			85.8	86.2	85.3
	✓	✓		86.5	86.8	86.1
	✓		✓	**87.9**	**88.3**	**87.2**

Experiment 3: To verify the influence of integrating DCNNs with GA on DR classification performance, on the basis of Experiment 2, three improved DCNNs were selected with multi-resolution images as input. Different DCNNs

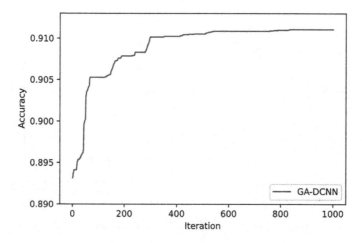

Fig. 4. Iteration curve of the GA-DCNN.

have different classification effects on the same image. The reason is that different DCNNs pay different attention to each pixel in the image. It can be seen that the fusion of the results of different DCNNs can play a complementary role. Therefore, the strategy of integrating DCNNs with GA is used to synthesize the classification results of the three DCNNs, and the iteration curve is shown in Fig. 4.

Table 6. Comparison of results between GA-DCNN and the improved DCNN model

Methods	Accuracy (%)	Sensitivity (%)	Specificity (%)
DenseNet201+SPP+GCA-SA	87.5	87.9	87.2
MobileNet+SPP+GCA-SA	89.1	88.6	87.3
InceptionV3+SPP+GCA-SA	87.9	88.3	87.2
GA-DCNN(Ours)	**91.2**	**91.8**	**90.7**

According to the results in Table 6, the proposed strategy of integrating DCNNs with GA makes the accuracy, specificity and sensitivity of the test dataset outperform the results of using a single DCNN model. In the same experimental setting, Fig. 5 shows that the AUC value of the GA-DCNN model is also higher than that of a single DCNN. The above results show that integrating DCNNs with GA can effectively compensate for the classification errors of the three DCNN models by iteratively optimizing the classification results of the three DCNNs. Therefore, compared with the single DCNN model, the proposed ensemble strategy can better improve the classification performance of the overall model.

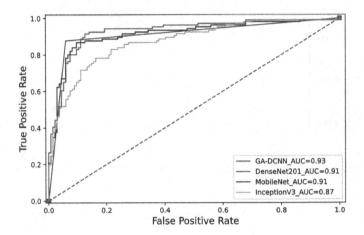

Fig. 5. ROC curves of GA-DCNN and improved DCNNs.

Table 7. Comparison of GA-DCNN and classical models

Methods	Accuracy (%)	Sensitivity (%)	Specificity (%)
VGG16	77.4	65.1	87.3
VGG19	80.4	67.9	90.3
ResNet50	70.4	40.6	94.0
ResNet152	67.9	32.1	96.3
DenseNet121	84.5	78.3	89.6
DenseNet169	87.0	82.1	91.0
InceptionResV2	82.4	78.3	85.8
MobileNetV2	87.5	87.5	88.8
NasNetMobile	75.8	54.7	92.5
Bagging	87.8	88.1	87.3
Adaboost	88.9	89.2	88.2
GA-DCNN(Ours)	**91.2**	**91.8**	**90.7**

Finally, in the same experimental environment, GA-DCNN is compared with classical CNN models. As can be seen from Table 7, GA-DCNN outperforms other classical models in accuracy, sensitivity and specificity. Therefore, the proposed GA-DCNN model is beneficial to make up for the shortcomings of a single model, integrates the decisions of multiple models, and improves the performance of DR classification.

5 Conclusion

This paper proposes a diabetic retinopathy classification model GA-DCNN based on the strategy of integrating DCNNs with GA. First, the fundus image features are highlighted by extracting the green channel and adaptive equalization in the preprocessing stage, and then the fine-grained lesion information of DR is learned by combing the DCNN and the GCA-SA modules. Finally, the strategy of integrating DCNNs with GA is used to make comprehensive decisions on the three DCNNs, which further improves the performance of DR classification. The experimental results demonstrate the effectiveness of the GA-DCNN model in DR detection and classification tasks. Furthermore, future research will be focused on applying GA-DCNN to the multi-classification task of DR and further optimizing the model performance by capturing the inter-class similarity of DR.

References

1. Silberman, N., Ahlrich, K., Fergus, R., et al.: Case for automated detection of diabetic retinopathy. In: Artificial Intelligence for Development, Papers from the AAAI Spring Symposium, pp. 1–10 (2010)
2. Saeedi, P., Salpea, P., Karuranga, S., et al.: Mortality attributable to diabetes in 20–79 years old adults, 2019 estimates: results from the international diabetes federation diabetes atlas, 9th edition. Diabetes Res. Clin. Pract. 1–7 (2020)
3. Sabanayagam, C., Banu, R., Chee, M.L., et al.: Incidence and progression of diabetic retinopathy: a systematic review. Lancet Diabetes Endocrinol. 7(2), 140–149 (2019)
4. Zhang, F.J., Li, J.M., Liu, Q.P.: Pathogenesis and potential treatment of diabetic retinopathy. Recent Adv. Ophthalmol. 40(7), 677–685 (2020)
5. Lecun, Y., Bengio, Y., Hinton, G.: Deep learning. Nature 521, 436–444 (2015)
6. Chan, H., Hadjiiski, L.M., Samala, R.K.: Computer-aided diagnosis in the era of deep learning. Med. Phys. 47(5), 218–227 (2020)
7. Al-Antary, M.T., Arafa, Y.: Multi-scale attention network for diabetic retinopathy classification. IEEE Access 9, 54190–54200 (2021)
8. Krause, J., Gulshan, V., Rahimy, E., et al.: Grader variability and the importance of reference standards for evaluating machine learning models for diabetic retinopathy. Ophthalmology 125(8), 1264–1272 (2018)
9. Qomariah, D. U. N., Tjandrasa, H., Fatichah, C.: Classification of diabetic retinopathy and normal retinal images using CNN and SVM. In: 2019 12th International Conference on Information Communication Technology and System, pp. 152–157 (2019)
10. Yang, Y., Shang, F., Wu, B., et al.: Robust collaborative learning of patch-level and image-level annotations for diabetic retinopathy grading from fundus image. IEEE Trans. Cybern. 99, 1–11 (2021)
11. Kamran, S. A., Tavakkoli, A., Zuckerbrod, S. L.: Improving robustness using joint attention network for detecting retinal degeneration from optical coherence tomography images. In: 2020 IEEE International Conference on Image Processing, pp. 2476–2480 (2020)
12. Ding, Y.F., Wen, S.G., Xie, J.Y., et al.: Weakly supervised attention pyramid convolutional neural network for fine-grained visual classification (2020)

13. He, K., Zhang, X., Ren, S., et al.: Spatial pyramid pooling in deep convolutional networks for visual recognition. IEEE Trans. Pattern Anal. Mach. Intell. **37**(9), 1904–1916 (2014)
14. Xia, H.Y., Lan, Y., Song, S.X., et al.: A multi-scale segmentation-to-classification network for tiny microaneurysm detection in fundus images. Knowl.-Based Syst. (2021)
15. Ramya, J., Rajakumar, M.P., Maheswari, B.U.: Deep CNN with hybrid binary local search and particle swarm optimizer for exudates classification from fundus images. J. Digit. Imaging **35**, 56–67 (2022)
16. Eftekheri, N., Masoudi, M., Pourreza, H., et al.: Microaneurysm detection in fundus images using a two-step convolutional neural network. Biomed. Eng. Online **18**(1), 1–16 (2019)
17. Gulshan, V., Peng, L., Coram, M., et al.: Development and validation of a deep learning algorithm for detection of diabetic retinopathy in retinal fundus photographs. JAMA **316**(22), 2402–2410 (2016)
18. Lin, Z., Guo, R., Wang, Y., et al.: A framework for identifying diabetic retinopathy based on anti-noise detection and attention-based fusion. In: International Conference on Medical Image Computing and Computer-assisted Intervention, pp. 74–82 (2018)
19. He, A., Li, T., Li, N., et al.: CABNet: category attention block for imbalanced diabetic retinopathy grading. IEEE Trans. Med. Imaging **40**(1), 143–153 (2021)
20. Nneji, G.U., Cai, J.Y., Deng, J.H., et al.: Identification of diabetic retinopathy using weighted fusion deep learning based on dual-channel fundus scans. Diagnostics **12**(2), 1–19 (2022)
21. Li, X., Jiang, Y.C., Zhang, J.D., et al.: Lesion-attention pyramid network for diabetic retinopathy grading. Artif. Intell. Med. **126**, 1–10 (2022)
22. Zhao, L.M., Wang, J., Li, X., et al.: On the connection of deep fusion to ensembling. CoRR (2016)
23. Gao, J.F., Sehrish, Q., Zhang, J.M., et al.: Ensemble framework of deep CNNs for diabetic retinopathy detection. Comput. Intell. Neurosci. (2020)
24. Zhuang, A., Xuan, H., Yuan, F., et al.: DR-IIXRN: detection algorithm of diabetic retinopathy based on deep ensemble learning and attention mechanism. Front. Neuroinform. **15**, 1–16 (2021)
25. Woo, S., Park, J., Lee, J.Y., et al.: CBAM: convolutional block attention module. In: European Conference on Computer Vision, pp. 3–19 (2018)
26. Holland, J.H.: Genetic algorithms and adaptation. In: Selfridge, O.G., Rissland, E.L., Arbib, M.A. (eds.) Adaptive Control of Ill-Defined Systems. NATO Conference Series, vol. 16, pp. 317–333. Springer, Boston, MA (1984). https://doi.org/10.1007/978-1-4684-8941-5_21
27. Dietterich, T.G.: Machine learning research: four current directions. AI Mag. **18**(4), 97–136 (1997)
28. Asia-pacific Society of Ophthalmology big Data Competition. https://tian-chi.aliyun.com/competition/entrance/531929/introduction. Accessed 4 Oct 2021

The Convolutional Neural Network Combing Feature-Aligned and Attention Pyramid for Fine-Grained Visual Classification

Enhui Shi🆔 and Ming Yang[✉]

School of Computer and Electronic Information/School of Artificial Intelligence,
Nanjing Normal University, Nanjing 210023, China
myang@njnu.edu.cn

Abstract. Fine-grained visual classification (FGVC) heavily relies on partial localization and part-based discriminant feature learning. The current methods mainly focus on extracting information from high-level features, while ignoring the influence of low-level features on FGVC. Based on this, this paper integrates low-level detailed information and high-level semantic information to improve the model performance by enhancing the feature representation and accurately locating the discriminant part. This paper proposes an end-to-end convolutional neural network combining feature-aligned and attention pyramid (FAAP-CNN), which consists of three main modules: 1) Feature pyramid: transmits high-level semantic information in a top-down path. Meanwhile, the semantic gap and information loss in information transmission are reduced through feature alignment and feature selection, and the integrity and reliability of high-level feature information are maintained. 2) Attention pyramid: pass the detailed information of low-level features in a bottom-up path to enhance the feature representation; 3) ROI feature refinement: dropblock and zoom-in are used for feature refinement to effectively eliminate background noise. The experimental results on three publicly FGVC datasets show that FAAP-CNN has excellent performance.

Keywords: Fine-grained visual classification · Multi-scale feature fusion · Attention Pyramid

1 Introduction

As a significant basic topic in computer vision, fine-grained visual classification (FGVC) has attracted the attention of many scholars, and has been widely used in many fields such as person re-identification [1], retail commodity recognition [2]. Compared with traditional classification, FGVC aims to divide category into different subcategories [3]. Because of the large intra-class variance and small inter-class variance of fine-grained images, it is much more difficult to classify fine-grained images than traditional classification tasks.

E. Chen et al. (Eds.): BigData 2023, CCIS 2005, pp. 61–75, 2023.
https://doi.org/10.1007/978-981-99-8979-9_5

In order to address the above challenges, researchers have dealt with the problems existing in FGVC from different aspects. The current FGVC methods can be roughly divided into two categories: 1) end-to-end feature encoding [4]; 2) regional localization [5]. The former is to capture more subtle features by calculating higher-order information, while the latter is to locate discriminant regions by attention mechanisms and deep filters [6].

Although promising results have been reported in the above studies, further improvements are limited by low-level information on CNN. Ding et al. [7] showed that the low-level information of CNN (e.g., color, edge information) is indeed essential in FGVC tasks. The deep layers of CNN have a relatively large receptive field and strong representation ability of semantic information, but the feature map resolution is low and the representation ability of spatial geometric features is weak [8]. It would make the detailed information of small distinguishing regions lost inevitably. In contrast to the deep layer, the low layer is rich in feature space information but lacking in semantic information. For fine-grained images, the small receptive field of low-level features allows it to detect more subtle parts such as the shape of claws. The high-level features of the receptive field can detect larger parts such as the bird's head or even the whole bird. These are all useful information for FGVC, which can effectively reflect the intra-class differences of fine-grained images.

Therefore, the feature pyramid structure is introduced to effectively extract and fuse multi-scale features, integrate high-level semantics and low-level details, and improve the performance of FGVC. Specifically, this paper designs a novel FGVC model, which can deal with semantic gap problems in multi-scale feature fusion, locate discriminant feature regions accurately, and reduce the impact of background noise on model accuracy availably. The main contributions can be summarized as follows:

1) A new feature pyramid structure is constructed in this paper, which can solve with the semantic gap in the fusion process of adjacent features, and effectively reduce the information loss caused by sharp dimensionality reduction.
2) A new attention structure is designed in this paper, which can focus on the object region, reduce the impact of background noise on classification performance, and enable the model to locate the discriminant region precisely.
3) Feature pyramid and attention pyramid constitute a new dual-path architecture, top-down feature path learning enhances high-level semantic information, bottom-up attention path learning low-level detailed representation, and the dual path structure works together to improve the model accuracy.
4) Three general fine-grained datasets (CUB-200-2011 [9], Stanford Cars [10], and FGVC-Aircraft [11]) are conducted for experimental verification. Ablation studies and visualization are performed to further verify the model performance. The results show that this model can improve the accuracy of FGVC.

The rest of the paper is organized as follows: Sect. 2 summarizes the related work of FGVC. Section 3 introduces the proposed model framework in detail. Section 4 discusses the experimental details, the final results and visualization analysis. And the last part shows the conclusion of this paper.

2 Related Work

The work of this paper is closely related to multi-scale features and attention for FGVC. Therefore, the following introduces these two types approaches.

2.1 Methods Using Multi-scale Information

Multi-scale information is accustomed to capturing features of different levels, providing a richer and more diversified image representation, improving the classification performance, and enhancing the robustness of model. Therefore, it is widely used in many fields such as object detection [12] and image classification [13]. Lin et al. [14] proposed a feature pyramid with a top-down and skip connection architecture, so that there is rich semantic information at all levels. Xu et al. [15] studied the characteristics of birds at different scales by using the pyramid to improve the accurate identification in the natural environment. Jiang et al. [16] adopted multi-scale fusion for the same type of features and multi-view fusion for different types of features to make the model learn food features from different granularity. Ding et al. [7] used FPN [14] and attention mechanism to effectively combine high-level feature semantic information with low-level feature details to improve FGVC. In this paper, according to the references [7] and [17], a new feature pyramid structure is designed, which can effectively reduce the information loss in the process of transferring semantic information from high-level to low-level, and improve the classification performance of model.

2.2 Methods Using Attention Mechanisms

Since the attention mechanism has the advantages of increasing the interpretability of model and enhancing the model's ability to learn features. Therefore, it is widely used in semantic segmentation [18], visual question answering [19], image classification [20] and so on. Fu et al. [21] and Zheng et al. [22] first applied the attention mechanism to FGVC, effectively improving the precision of FGVC. Zheng et al. [23] proposed a multi-level attention model to obtain object-level and partial-level attention respectively. Han et al. [24] took SENet [25] as a partial positioning network to enhance the feature representation capability of FGVC. Ding et al. [7] improved on CBAM [26] and used attention mechanism to transmit low-level feature information to guide feature refinement in stage II. With reference to [27], this paper improves the attention structure and enhances multi-scale features by combining space and channel levels: 1) Adopt spatial information to guide feature refinement in stage II; 2) Use channel information to transfer low-level details to high-level features.

3 The Convolutional Neural Network Combing Feature-Aligned and Attention Pyramid

In this section, the convolutional neural network combing feature-aligned and attention pyramid (FAAP-CNN) is designed. FAAP-CNN is an end-to-end two-stage network, the first stage adopts the complete image as input, the second stage uses the cropped and enlarged local feature information as input. Both stages share the same network structure and have the same parameters. The model framework is shown in Fig. 1. Firstly, an image is input to generate a feature pyramid and an attention pyramid via CNN. Secondly, the spatial attention pyramid obtained from the original image of stage I is used to guide the feature refinement of stage II, and the ROI pyramid is constructed by generating the region of interest through the region proposal generator with NMS operation. Specifically, the feature refinement is achieved by dropblock to remove the most discriminative regions of the low-level features, merging and enlarging all ROI regions. Finally, the refined feature information is adopted as input to carry out in stage II. The final classification result is the average of stage I and stage II.

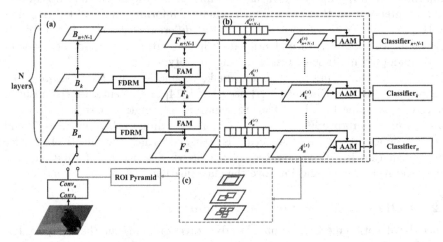

Fig. 1. The framework of FAAP-CNN. (a) feature and attention dual pathway; (b) attention pyramid; (c) ROI pyramid. In this figture, the feature maps are represented as blue frames and the spatial/channel attention are indicated by green outlines. (Color figure online)

3.1 Bottom-Up Multi-scale Feature Module

Motivataion. The purpose of this paper is to integrate high-level semantics and low-level details to improve the performance of FGVC. Therefore, a bottom-up multi-scale feature module is introduced to transfer high-level semantic information to low-level features. Specifically, the output feature maps of each convolutional block in the backbone network are denoted as B_1, B_2,..., B_l, where l represents the number of blocks. FPN uses feature reduction and feature fusion to generate new feature maps $\{F_n, F_{n+1}, \ldots, F_{n+N-1}\}$ $(1 \leq n \leq n + N - 1 \leq l)$, the comparison between the original FPN and the FPN proposed in this paper is shown in Fig. 2. $B_k \rightarrow F_k$ is used to maintain backbone information, and $F_{k+1} \rightarrow F_k$ is used to convey high-level semantic information from the top down. FAM and FDRM are described in detail below.

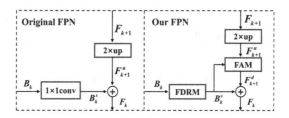

Fig. 2. Comparison between the original FPN and our FPN. B_k^s represents the features after dimensionality reduction; F_{k+1}^u indicates the features after upsampling; F_k indicates the fusion feature of F_{k+1}^u and B_k^s. $\oplus$ denotes broadcasting addition.

Feature Alignment. Due to the use of upsampling, two adjacent features in FPN are fused by adding elements, ignoring the semantic gap between feature maps caused by different depth, which will produce wrong prediction of the boundary of the object and affect the subsequent localization and recognition. Therefore, this paper refers to [17] and introduces deformable convolution (DCN) to adjust the spatial position information of F_{k+1}, achieving feature alignment between B_k and F_{k+1}. The details are shown in Fig. 3. The feature alignment can be expressed as:

$$\Delta_i = f_o\left(\left[B_k^s, F_{k+1}^u\right]\right) \tag{1}$$

$$F_{k+1}^d = f_a\left(F_{k+1}^u, \Delta_i\right) \tag{2}$$

where [ ] denotes the cascade of B_k^s and F_{k+1}^u, representing the spatial difference between the two features, and f_o indicates the function of learning the offset from 2D coordinate system; f_a denotes the function that adjusts the F_{k+1}^u based on the learned offset. Both functions are implemented using DCN [27].

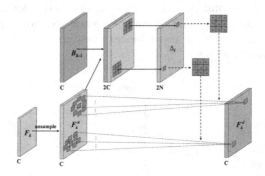

Fig. 3. The description of feature alignment. N represents N sampling points, and for 3×3 convolution, $N = 3 \times 3 = 9$. For feature maps with [C, h, w], the learned offset is [2N, h, w]. Each value in Δ_i denotes the horizontal or vertical offset of the corresponding feature position.

This paper reviews DCN briefly. DCN is a geometric transformation of the convolution kernel that enables the model's attention sampling area to be more concentrated on the target itself. Firstly, an input feature map $d_i \in R^{H_i \times W_i}$ and a 3×3 convolution layer are defined. The feature of a position $\hat{x}_p$ in feature map can be expressed by the following formula:

$$\hat{x}_p = \sum_{n=1}^{N} w_n \cdot (x_p + p_n) \tag{3}$$

where $N = 3 \times 3 = 9$, w_n and $p_n \in \{(-1,-1),(-1,0),\ldots,(1,1)\}$ refers to the weight and pre-specified offset of the N_{th} convolution position, respectively. From standard convolution to deformable convolution, an offset-guided convolution kernel is required to sample the position of the deformable convolution. Deformable convolution attempts to adaptively learn additional offsets $\{\Delta p_1, \Delta p_2, \ldots, \Delta p_N\}$, the above equation can be re-expressed as:

$$\hat{x}_p = \sum_{n=1}^{N} w_n \cdot (x_p + p_n + \Delta p_n) \tag{4}$$

In this paper, when using deformable convolution in FPN, the connection of B_k and F_{k+1}^u is used to learn the offsets of feature maps, and the spatial position information of current features is adjusted by formula 1 and 2 to achieve similar focus areas for adjacent features, thereby achieving feature alignment to address the semantic gap problem.

Feature Selection. In addition to semantic gap, FPN also has the problem of information loss. The horizontal connection of 1×1 convolution layer can produce features with the same dimension, but due to the sharp dimensionality reduction, the extracted feature maps, especially the high-level feature maps, suffer serious information loss. Therefore, this article proposes a feature dimensionality reduction module, which uses channel weight as an important indicator for feature maps, and sets different sampling ratios for different feature maps to highlight important information and reduce redundant information. By recalibrating the feature map, feature dimensionality reduction is achieved. The FDRM is illustrated in Fig. 4.

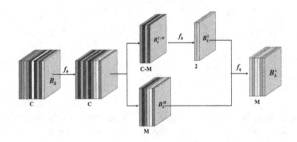

Fig. 4. The description of feature selection.

Firstly, the information of B_k is extracted by global average pooling and global maximum pooling respectively, and the two are added together, and the channel weights corresponding to the feature maps are obtained by sigmoid. Secondly, the feature maps are recalibrated according to the channel weight information. Select the first M feature maps with high values, and the rest feature maps extract global information through global maximum pooling and global average pooling. Finally, the 1×1 convolution is used to reduce the dimensions of the cascaded B_k^M and B_k^2, and the final feature B_k^s is obtained. In general, the FDRM can be formulated as:

$$B_k^2 = f_b \left[\text{AvgPool} \left(B_k^{C-M} \right) ; \text{MaxPool} \left(B_k^{C-M} \right) \right] \tag{5}$$

$$B_k^s = f_q \left(\left[B_k^M, B_k^2 \right] \right) \tag{6}$$

where f_q represents 1×1 convolution, B_k^M represents the first M feature maps, and B_k^{C-M} represents the remaining feature maps.

The feature maps obtained from the FDRM module not only include important feature information, but also reduce the impact of redundant information in the current layer on subsequent feature fusion, which improves the positioning accuracy of the attention module to a certain extent.

3.2   Top-Down Attention Module

Motivation. The purpose of this module is to enhance the detailed representation of features. Since the feature fusion operation introduces deep semantic information into low-level features, it improves the discriminant ability of low-level features and helps to identify more subtle parts of objects. Due to the lack of sufficient semantic information to distinguish objects, low-level features pay too much attention to noise, which will cause attention deviation. Due to the lack of detailed information constraints, deep features will lead to the divergence of attention. Because of the introduction of deep semantic information, the background noise is effectively suppressed, and the low-level positioning information provides constraints, which limits the attention to a certain range. Therefore, this paper uses the attention mechanism after FP, which is helpful to improve the performance of FGVC.

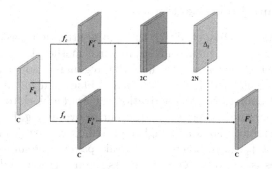

Fig. 5. The presentation of attention mechanism. The specific process of this module is much the same as the feature alignment.

Attention Mechanism. Because the hybrid attention can pay more attention to the spatial information and semantic information of object, this paper chooses the hybrid attention enhancement feature. However, the features enhanced by channel attention and spatial attention will be different to some extent, and direct fusion and addition operation may destroy the original feature representation. Therefore, this article introduces DCN to adjust features based on hybrid attention. The specific details are shown in Fig. 5.

First of all, the spatial enhanced features and channel enhanced features are aligned along the channel axis, and then 1×1 convolution is used to reduce dimension, and the same position feature values of all channels are added. The single feature map obtained in this way can better represent the entire image information, and the offset learned from features containing a lot of information can more accurately describe the spatial position of the target object. Semantic information has been integrated through 1×1 convolution, so that the position of the object covered by DCNv2 will be more accurate, so that the features after spatial enhancement can be fixed in the target object region after adjustment, and the positioning can be accurate. The detailed implementation of this module is shown below.

$$A_k^c = \sigma \left(W_2 \cdot \text{ReLU} \left(W_1 \cdot \text{AvgPool} \left(F_k \right) \right) + W_2 \cdot \text{ReLU} \left(W_1 \cdot \text{MaxPool} \left(F_k \right) \right) \right) \tag{7}$$

$$A_k^s = \sigma \left(f_b \left(\left[\text{AvgPool} \left(F_k \right); \text{MaxPool} \left(F_k \right) \right] \right) \right) \tag{8}$$

$$F_k^c = F_k \otimes A_k^c \quad F_k^s = F_k \otimes A_k^s \tag{9}$$

$$\Delta_i = f_o \left(\left[F_k^s, F_k^c \right] \right) \tag{10}$$

$$F_k' = f_a \left(F_k^s, \Delta i \right) \tag{11}$$

where A_k^c represents channel attention weight of feature F_k, A_k^s represents spatial attention weight of feature F_k, F_k^c and F_k^s represent channel enhancement feature and space enhancement feature respectively. f_c and f_s in Fig. 5 are shown in formulas 8 and 9 respectively. f_o and f_a have the same meaning as in FAM, meaning learning from the feature offset and adjusting according to the offset, respectively.

3.3   ROI Feature Refinement

Inspired by the idea of Ding et al. [7], the spatial attention mask is used as the anchor score, and the non-maximum suppression method is adopted to reduce the redundant anchor frames, and the final ROI region is denoted as $R_i(i = n, n + 1, \ldots, n + N - 1)$. This module mainly uses dropblock and zoomin operations to refine features. Dropblock sets the activation values of a block to 0, reducing the model's dependence on that region and preventing overfitting. For the dropblock operation, smaller block sizes and higher probabilities may lead the model to learn more robust features. Therefore, in this paper, the smaller region with the highest scores is set to 0. Zoomin adjusts the image size and target size by zooming in on the image. In the ROI feature refinement module, zoomin is used to adjust the ROI size of the input feature map. Figure 6 shows the process of guiding feature refinement.

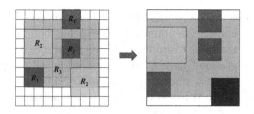

Fig. 6. Feature refinement.

First, all R_i are mapped into the same space, and dropblock operation is performed on the ROI feature with the highest anchor box score, and its feature value is set to 0 to prevent overfitting. Then, find out the minimum horizontal coordinate x_{min}, maximum horizontal coordinate x_{max}, minimum vertical coordinate y_{min} and maximum vertical coordinate y_{max} of all anchor frames, and compose them into four coordinates as the range of the new frame. Finally, enlarge it to the same size as the original space. Up to this point, the feature refinement operation is implemented. This operation can effectively remove the redundant background information and enhance the feature learning ability of the model.

4 Experimental Results and Analysis

Experimental validation is performed on three FGVC public datasets, including CUB-200-2011, Stanford Cars, and FGVC-Aircraft. The details of the datasets are shown in Table 1.

Table 1. The information of datasets

Datasets	Classes	Training	Testing
CUB-200-2011	200	5994	5794
Stanford-Cars	196	8144	8041
FGVC-Aircraft	100	6667	3333

In the analysis of experimental results, the three datasets are replaced by birds, cars and airs respectively.

4.1 Model Implementation Details

The FAAP-CNN is implemented on ResNet50 pretrained on ImageNet. Specifically, the final output features of the ResNet50's last three residual blocks, conv 3, conv 4, and conv 5, are utilized to construct a feature pyramid, denoted as B_3, B_4, and B_5. And the feature refinement is carried out on B_3. Input image is resized to 448×448 pixels. Additionally, pyramid levels are assigned anchors with scales of 64, 128, 256 and 1:1 ratio for each anchor. The most active 5, 3, and 1 anchors are chosen for potential refinement. The open-source PyTorch framework is used for model training, which is conducted on a single TITAN Xp GPU. Stochastic gradient descent method is adopted for optimization, with momentum set to 0.9 and a mini-batch size of 16. The initial learning rate is 0.001, which diminishes to 0 based on a cosine annealing schedule. All models are trained for a total of 100 epochs.

4.2 Comparison with State-of-the-art Methods

Table 2 shows the performance evaluation of the above three datasets. Each column contains nine representative FGVC methods. This paper presents the results of model based on ResNet50 as the backbone network. Compared with the above methods, the FAAP-CNN proposed in this paper has a certain degree of performance improvement on the three datasets. The comparison results can be summarized as follows: On birds dataset, the two-stage FAAP-CNN model proposed in this paper achieves 88.9% accuracy. As can be seen from Table 2, Du et al. and PMG adopted progressive training strategies to learn complementary attributes of different granularity, achieving an accuracy of 89.9% and 89.6%, respectively. On the cars dataset, FAAP-CNN outperforms Du et al. by 0.7% with an accuracy of 96.1%. On airs dataset, FAAP-CNN achieves the best accuracy of 94.9%. Compared with Du et al., the accuracy is improved by 0.8%. The comparison of the above experimental results can demonstrate the effectiveness of the proposed model.

Table 2. Comparison results of three datasets

Methods	Base	Image resolution	Birds (%)	Cars (%)	Airs (%)
CIN (AAAI 20 [28])	ResNet50	448 × 448	87.5	94.1	92.8
MC-Loss (TIP 20 [29])	ResNet50	448 × 448	87.3	93.7	92.6
PMG (ECCV 20 [30])	ResNet50	448 × 448	89.6	95.1	93.4
AP-CNN (TIP 21 [7])	ResNet50	448 × 448	88.4	95.4	94.1
DTRG (TIP 22 [31])	ResNet50	448 × 448	88.8	95.2	94.1
MSAC (ICME 21 [32])	ResNet50	448 × 448	88.3	94.6	92.9
Du et al. (TPAMI 21 [33])	ResNet50	448 × 448	**89.9**	95.4	94.1
Song et al. (TPAMI 22 [34])	ResNet50	448 × 448	86.2	93.6	91.4
iSICE (CVPR 23 [4])	ResNet50	448 × 448	85.9	93.5	92.7
FAAP-CNN (one stage)	ResNet50	448 × 448	87.9	94.3	93.0
FAAP-CNN (two stage)	ResNet50	448 × 448	88.9	**96.1**	**94.9**

In general, the advantages of FAAP-CNN proposed in this paper lie in two parts: 1) By establishing a dual path to integrate high-level semantics and low-level details. The integrity and reliability of high-level feature information are maintained through feature alignment and feature selection, and the accuracy of classification and location is improved. 2) Through the ROI-guided feature refinement stage, the background noise is further eliminated, and the feature learning ability of the model in stage II is enhanced, which is conducive to the improvement of performance.

Table 3. Comparison results of different modules

Methods	Base	Accuracy (%)
baseline	ResNet50	84.1
FP	ResNet50	86.6
FP + AP	ResNet50	87.2
FP + AP + FDRM	ResNet50	87.4
FP + AP + FAM	ResNet50	87.5
FP + AP + AAM (adjust spatial features)	ResNet50	87.5
FP + AP + AAM (adjust channel features)	ResNet50	87.5
FP + AP + FDRM + FAM	ResNet50	87.7
FP + AP + FDRM + FAM + AAM	ResNet50	87.9

Table 4. Detailed information for the three modules. A: feature pyramid. B: attention pyramid. C: ROI feature refinement

Methods	Base	GFlops	Params	Time
A	ResNet50	325.12	27.92M	252.2 s/epoch
A + B	ResNet50	341.97	28.20M	315.3 s/epoch
A + B + C	ResNet50	565.75	28.20M	658.1 s/epoch

4.3 Ablation Studies

In this paper, some ablation experiments are conducted to analyze the contribution of each module. The following experiments are conducted on the birds dataset, all using ResNet50 as the backbone network. Table 3 shows the contributions of three modules: feature alignment, feature selection, and attention alignment. Table 4 represents the parameter details of three modules. Since this paper focuses on changes made to the AP-CNN model, a comparative analysis of the computational complexity, params, and training efficiency of AP-CNN and FAAP-CNN is conducted. The specific results are shown in Table 5.

As shown in Table 3, FP leads obvious performance improvement compared with baseline, and AP further improves accuracy by enhancing the correlation between features, indicating that the structure of the dual path is very meaningful for FGVC. With the superposition of FAM, FDRM and AAM, the performance of the model is gradually improved, which shows the significance of the work proposed in this paper. When AAM is added separately, no matter the spatial feature or channel feature is adjusted, the performance of the model can be improved to some extent, which indicates that the module has the same binding force on the two features, and both of them can focus on the target object

Table 5. Comparative analysis of FAAP-CNN and AP-CNN

Methods	Base	GFlops	Params	Time
AP-CNN	ResNet50	31.93	27.96M	389 s/epoch
FAAP-CNN	ResNet50	565.75	28.20M	658.1 s/epoch

region. Table 4 shows the details of the calculation cost, parameter and time spent on a single GPU for the feature module, attention module and ROI refinement module. It can be seen that the three parameters are gradually increasing. As can be seen from Table 5, the computational complexity of FAAP-CNN is significantly higher than that of AP-CNN. Generally speaking, the higher the computational complexity, the longer the required computing time. Therefore, FAAP-CNN performs poorly in training speed, which is almost twice that of AP-CNN. In terms of model complexity, the number of parameters in FAAP-CNN is slightly higher than that in AP-CNN, indicating that the improved module of the FAAP-CNN model does not add a lot of parameters to the AP-CNN model, and the improvement in the module is worth considering. Analysis shows that FAAP-CNN consumes a lot of computational resources during training, and the model takes longer to converge. In future work, we will further optimize the structure of the FAAP-CNN model to reduce unnecessary calculations and reduce the computational complexity of the model.

4.4 Visualization

Fig. 7 shows the ROI pyramid obtained from FAAP-CNN. For each dataset, this paper selects two test images and uses red, blue, and green boxes to represent

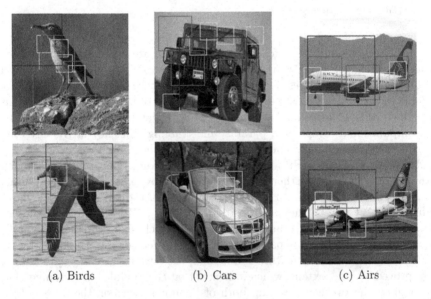

(a) Birds   (b) Cars   (c) Airs

Fig. 7. Visualization of ROI pyramid.

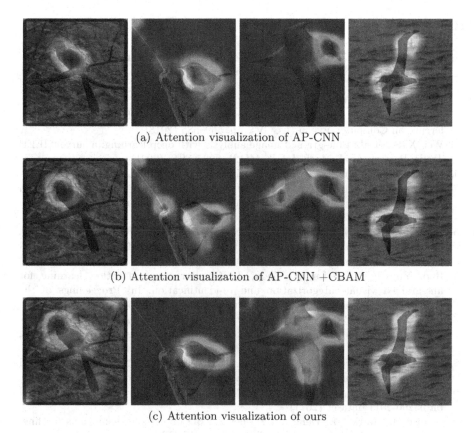

(a) Attention visualization of AP-CNN

(b) Attention visualization of AP-CNN +CBAM

(c) Attention visualization of ours

Fig. 8. Visualization of attention.

regions of interst. As can be intuitively observed in Fig. 7, local regions from different levels can focus the model on more subtle and discriminative parts. Figure 8 shows the attention visualization of the model presented in this article. Compared with the original attention pyramid [7] (AP-CNN) and the attention pyramid integrated with CBAM, it can be intuitively seen that the attention structure proposed in this paper can focus on more subtle and discriminative parts, and its performance is relatively superior.

5   Conclusion

In this paper, a convolutional neural network combining feature-aligned and attention pyramid is proposed for FGVC. This method uses dual paths to integrate high-level and low-level information to enrich feature representation at each level. The feature refinement in stage II further eliminates the impact of background noise on classification and improves the classification performance of the model. Experimental results show that the proposed method performs well on three public datasets.

References

1. Huang, M., Hou, C., Yang, Q., Wang, Z.: Reasoning and tuning: graph attention network for occluded person re-identification. IEEE Trans. Image Process. **32**, 1568–1582 (2023)
2. Follmann, P., Bottger, T., Hartinger, P., Konig, R., Ulrich, M.: MVTec D2S: densely segmented supermarket dataset. In: Proceedings of the European Conference on Computer Vision (ECCV), pp. 569–585 (2018)
3. Wei, X.-S., et al.: Fine-grained image analysis with deep learning: a survey. IEEE Trans. Pattern Anal. Mach. Intell. 44(12), 8927–8948 (2021)
4. Rahman, S., Koniusz, P., Wang, L., Zhou, L., Moghadam, P., Sun, C.: Learning partial correlation based deep visual representation for image classification. In: Proceedings of the IEEE/CVF Conference on Computer Vision and Pattern Recognition, pp. 6231–6240 (2023)
5. Huang, Z., Li, Y.: Interpretable and accurate fine-grained recognition via region grouping. In: Proceedings of the IEEE/CVF Conference on Computer Vision and Pattern Recognition, pp. 8662–8672 (2020)
6. Rao, Y., Chen, G., Lu, J., Zhou, J.: Counterfactual attention learning for fine-grained visual categorization and re-identification. In: Proceedings of the IEEE/CVF International Conference on Computer Vision, pp. 1025–1034 (2021)
7. Ding, Y., et al.: AP-CNN: weakly supervised attention pyramid convolutional neural network for fine-grained visual classification. IEEE Trans. Image Process. **30**, 2826–2836 (2021)
8. Luo, W., Li, Y., Urtasun, R., Zemel, R.: Understanding the effective receptive field in deep convolutional neural networks. Adv. Neural Inf. Process. Syst. **29** (2016)
9. Wah, C., Branson, S., Welinder, P., Perona, P., Belongie, S.: The caltech-ucsd birds-200-2011 dataset (2011)
10. Krause, J., Stark, M., Deng, J., Fei-Fei, L.: 3D object representations for fine-grained categorization. In: Proceedings of the IEEE International Conference on Computer Vision Workshops, pp. 554–561 (2013)
11. Maji, S., Rahtu, E., Kannala, J., Blaschko, M., Vedaldi, A.: Fine-grained visual classification of aircraft. arXiv preprint arXiv:1306.5151 (2013)
12. Liu, D., Liang, J., Geng, T., Loui, A., Zhou, T.: Tripartite feature enhanced pyramid network for dense prediction. IEEE Trans. Image Process. (2023)
13. Jin, Y., Liu, J., Chen, H., Duan, W., Cao, D., Pang, B.: MASKED-AP: attention pyramid convolutional neural network with mask for cervical cell classification. In: ICASSP 2023–2023 IEEE International Conference on Acoustics, Speech and Signal Processing (ICASSP), pp. 1–5. IEEE (2023)
14. Lin, T.-Y., Dollár, P., Girshick, R., He, K., Hariharan, B., Belongie, S.: Feature pyramid networks for object detection. In: Proceedings of the IEEE Conference on Computer Vision and Pattern Recognition, pp. 2117–2125 (2017)
15. Xu, X., Yang, C.-C., Xiao, Y., Kong, J.-L.: A fine-grained recognition neural network with high-order feature maps via graph-based embedding for natural bird diversity conservation. Int. J. Environ. Res. Public Health **20**(6), 4924 (2023)
16. Jiang, S., Min, W., Liu, L., Luo, Z.: Multi-scale multi-view deep feature aggregation for food recognition. IEEE Trans. Image Process. **29**, 265–276 (2019)
17. Huang, S., Lu, Z., Cheng, R., He, C.: FaPN: feature-aligned pyramid network for dense image prediction. In: Proceedings of the IEEE/CVF International Conference on Computer Vision, pp. 864–873 (2021)

18. Jia, Z., et al.: Event-based semantic segmentation with posterior attention. IEEE Trans. Image Process. **32**, 1829–1842 (2023)
19. Zhang, H., Li, R., Liu, L.: Multi-head attention fusion network for visual question answering. In: 2022 IEEE International Conference on Multimedia and Expo (ICME), pp. 1–6. IEEE (2022)
20. Xu, B., Zhang, W.: Selective scale cascade attention network for breast cancer histopathology image classification. In: ICASSP 2022–2022 IEEE International Conference on Acoustics, Speech and Signal Processing (ICASSP), pp. 1396–1400. IEEE (2022)
21. Fu, J., Zheng, H., Mei, T.: Look closer to see better: recurrent attention convolutional neural network for fine-grained image recognition. In: Proceedings of the IEEE Conference on Computer Vision and Pattern Recognition, pp. 4438–4446 (2017)
22. Zheng, H., Fu, J., Mei, T., Luo, J.: Learning multi-attention convolutional neural network for fine-grained image recognition. In: Proceedings of the IEEE International Conference on Computer Vision, pp. 5209–5217 (2017)
23. Zheng, H., Fu, J., Zha, Z.-J., Luo, J., Mei, T.: Learning rich part hierarchies with progressive attention networks for fine-grained image recognition. IEEE Trans. Image Process. **29**, 476–488 (2019)
24. Han, J., Yao, X., Cheng, G., Feng, X., Xu, D.: P-CNN: part-based convolutional neural networks for fine-grained visual categorization. IEEE Trans. Pattern Anal. Mach. Intell. **44**(2), 579–590 (2019)
25. Hu, J., Shen, L., Sun, G.: Squeeze-and-excitation networks. In: Proceedings of the IEEE Conference on Computer Vision and Pattern Recognition, pp. 7132–7141 (2018)
26. Woo, S., Park, J., Lee, J.-Y., Kweon, I.S.: CBAM: convolutional block attention module. In: Proceedings of the European Conference on Computer Vision (ECCV), pp. 3–19 (2018)
27. Zhu, X., Hu, H., Lin, S., Dai, J.: Deformable convnets v2: more deformable, better results (2018)
28. Gao, Y., Han, X., Wang, X., Huang, W., Scott, M.: Channel interaction networks for fine-grained image categorization. In: Proceedings of the AAAI Conference on Artificial Intelligence, vol. 34, no. 07, pp. 10 818–10 825 (2020)
29. Chang, D., et al.: The devil is in the channels: mutual-channel loss for fine-grained image classification. IEEE Trans. Image Process. **29**, 4683–4695 (2020)
30. Du, R., et al.: Fine-Grained Visual Classification via Progressive Multi-granularity Training of Jigsaw Patches. In: Vedaldi, A., Bischof, H., Brox, T., Frahm, J.-M. (eds.) ECCV 2020. LNCS, vol. 12365, pp. 153–168. Springer, Cham (2020). https://doi.org/10.1007/978-3-030-58565-5_10
31. Liu, K., Chen, K., Jia, K.: Convolutional fine-grained classification with self-supervised target relation regularization. IEEE Trans. Image Process. **31**, 5570–5584 (2022)
32. Hou, Y., Zhang, W., Zhou, D., Ge, H., Zhang, Q., Wei, X.: Multi-scale attention constraint network for fine-grained visual classification. In: 2021 IEEE International Conference on Multimedia and Expo (ICME), pp. 1–6. IEEE (2021)
33. Du, R., Xie, J., Ma, Z., Chang, D., Song, Y.-Z., Guo, J.: Progressive learning of category-consistent multi-granularity features for fine-grained visual classification. IEEE Trans. Pattern Anal. Mach. Intell. **44**(12), 9521–9535 (2021)
34. Song, Y., Sebe, N., Wang, W.: On the eigenvalues of global covariance pooling for fine-grained visual recognition. IEEE Trans. Pattern Anal. Mach. Intell. **45**(3), 3554–3566 (2022)

OCWYOLO: A Road Depression
Detection Method

Linmin Zhao, Bin Jiang[✉], Chao Yang, Qun Zhang, and Xiao Fang

College of Computer Science and Electronic Engineering, Hunan University,
Changsha 410082, Hunan, China
{xixizz,jiangbin,yangchaoedu,zqlll,fangxiao}@hnu.edu.cn

Abstract. In terms of road depression detection, there are four existing methods: (1) based on two-dimensional image processing methods, (2) three-dimensional point cloud modeling and segmentation based methods, (3) based on machine/deep learning methods, and (2) (3) hybrid methods. This article proposes a road depression detection algorithm based on Yolov7, which utilizes the Yolov7 network model to make different improvements: to make the localization more accurate, we use WIoU, which reduces the total loss and improves the AP to a certain extent. Due to the presence of some small objects in the dataset, small objective detection has always been a focus of object detection, so ODConv, which can effectively detect small objects, has been introduced. In the past, adding attention mechanisms directly added an attention component, which would make the model more complex. Therefore, we directly applied attention mechanisms to existing components without increasing the complexity of the model and achieving the goal of improving accuracy. In addition, we conducted a large number of experiments to verify the superiority of our model. We not only compare it on our road depression dataset but also conducted comparative experiments with the recent state-of-the-art models in the Yolo field on the COCO dataset to verify its universality. Compared to Yolov7, our model has improved by 14.9% on AP, while also comparing with the newly proposed Yolov8 model, we have improved by 2.0% on AP.

Keywords: Road depression detection · Yolov7 · Attention mechanism · Small objects

1 Introduction

A pothole is a structural road damage with considerable impact formed by the co-existence of water and traffic. Water seeps into the ground, weakening the soil under the pavement, and then traffic breaks the affected pavement, resulting in the loss of part of the pavement mass [1]. Potholes in roads cause significant inconvenience and pose a major threat to traffic safety. Road diseases such as cracks and potholes have a significant impact on the structure and service life of highways. According to The Pothole Facts, about one-third of the 33,000 traffic

© The Author(s), under exclusive license to Springer Nature Singapore Pte Ltd. 2023
E. Chen et al. (Eds.): BigData 2023, CCIS 2005, pp. 76–87, 2023.
https://doi.org/10.1007/978-981-99-8979-9_6

accidents in the U.S. are related to poor road conditions. Therefore, inspecting roads frequently and repairing potholes and depressions is necessary [2]. The naked eye directly identifies early inspection through full-time personnel and simple shooting equipment, which is undoubtedly very time-consuming and laborious, and the detection accuracy is not high.

In recent years, with the depth use of computers, computer vision technology has been widely used to obtain three-dimensional road information data and detect road potholes. The existing road pothole detection methods are mainly divided into four categories: (1) Classical two-dimensional image processing [3] on MATLAB, the road is divided into non-defective parts and defective parts based on the threshold of histogram shape, and then the texture of the pit shape is extracted based on the geometric characteristics of the defective area and compared with the surrounding non-defective area to determine whether it is a pothole area. (2) 3D point cloud modeling and segmentation-based [4] 3D point cloud points with accurate height information are captured during scanning by 3D laser scanners. Then mesh-based processing methods focus on a specific feature, such as potholes. (3) Machine/deep learning-based [5] takes the histogram-based texture metric as the feature of the image and establishes a nonlinear support vector machine to identify whether the target area is a pothole. and (4) the mixed [2] first transform the dense disparity map, extract the potential undamaged area from the transformed disparity map, and model the difference in the extracted area to obtain the modeling parallax map. By comparing the difference between the actual parallax map and the modeled parallax map, the location information of the pothole can be accurately detected.

However, because some potholes will be relatively small, we often want to be able to detect small potholes as early as possible to prevent them from gradually evolving into large depressions and making repair costs high, which involves small object detection. At present, small object detection methods emerge endlessly: SSD [6] proposes that in convolutional neural networks, the semantic information contained in the feature map of each output stage is different, and the semantic information output by the lower layer is more robust, so the lower output layer can be used to detect small objects, which can effectively improve the detection accuracy, which is the earliest feature pyramid structure. BiFormer [7] first filters out irrelevant key-value pairs in coarse-level areas and then applies fine-grained token-to-token attention in the remaining regions. YOLO-Z [8] improves the performance of the model detection of small objects to a certain extent by replacing the Backbone module with Resnet50 [9] and DenseNet [10] based on Yolov5 [11].

Overall, the main contributions of this work are the following three aspects:

(1) In order to achieve a more accurate localization effect, wise-iou is used to replace the original DIOU to optimize the loss function and improve the model's generalization ability.

(2) To improve the detection accuracy of small potholes, ODConv is used.

(3) In order to guide the object detection process and enhance the detection accuracy, coordinate attention mechanism is introduced without adding components.

2   Related Work

This section will briefly review the methods used in this area.

2.1   Object Detection Method

The object detection phase before 2012 is known as the "cold weapon era," when detection systems use classifiers to evaluate slices of different measured images. Since 2014, object detection has been divided into single-stage and two-stage. The two-stage detector mainly adds a proposed candidate box stage, according to the candidate box, to derive the final prediction box, such as R-CNN [12], run a segmentation algorithm to split an image into small pieces, and then run a classifier on these small blocks. Yolo [13] is a typical single-stage detector that constructs object detection as a regression problem into spatially separated bounding boxes and associated class probabilities. A single neural network predicts bounding boxes and class probabilities directly from the whole image in a single evaluation. Because the entire inspection pipeline is a single network, end-to-end optimization can be done directly on inspection performance.

Yolov7 [14] proposes several bag-of-freebies that can be used for training to improve accuracy by increasing the burden on training. However, the burden on inference does not increase, so the detection speed does not slow. At training time, a module is split into multiple identical or different module branches, and a multi-branch network is used to obtain better feature representation. When inference, multiple branches are combined into one equivalent module, reducing computation and parameters and improving the detection speed. Reference [15–19] shows that deeper networks can learn and converge efficiently by controlling the shortest and longest gradient paths. On this basis, Yolov7 proposed E-ELAN.

2.2   Intersection Over Union

The loss function for bounding box regression(BBR) is a critical factor in achieving object detection. If well-defined, this will bring significant performance gains to the model. With the improvement of the Yolo algorithm, different kinds of IoU have also emerged, such as CIoU, DIoU, EIoU, SIoU, etc. Most existing work assumes that the training data is of high quality and focuses on strengthening the fitting ability of BBR losses. If BBR is mindlessly enhanced on low-quality samples, it will specifically impact localization performance. Focal-EIoU [20] considered this problem but did not fully exploit the potential of non-monotonic focusing mechanisms(FM) because it used a static focusing mechanism. To solve this problem, a dynamic non-monotonic focusing mechanism loss method based on IoU is proposed and named Wise-IoU (WIoU) [21]. Dynamic non-monotonic frequency utilizes outliers instead of IoU to evaluate the quality of anchor frames and provides a sensible gradient gain distribution strategy. This strategy reduces the competitiveness of high-quality anchor frames while also reducing the harmful gradients generated by low-quality samples. This allows WIoU to focus on normal-quality anchor frames and improve the overall performance of the detector. Since it does not involve the aspect ratio calculation, the overall operating speed will be improved to a certain extent.

2.3   Dynamic Weight Networks

Learning a single static convolution kernel is a typical training paradigm in each convolutional layer of modern convolutional neural networks [22,23]. Reference [22,23] shows that learning a linear combination of n convolution kernels, with attention weights associated with the input, will significantly improve the accuracy of lightweight convolutional neural networks. Nevertheless, replacing ordinary convolution with this dynamic convolution will increase the number of convolution parameters by n times and only focus on the dynamic characteristics of one dimension (the number of convolution kernels) in the kernel space. Omni-dimensional Dynamic Convolution (ODConv) [24] utilizes a multidimensional attention mechanism (multi-headed SENet [25]) to learn four attention types (number of convolution kernels, convolution kernel size, number of input channels, number of output channels) in the kernel space in a parallel manner. In this way, ODConv with only one core can also achieve the learning effect of convolution with multiple cores and may even surpass it.

2.4   Attention Mechanism

Reference [26] applies attention mechanisms to computer vision for the first time. SENet captures the weight of each channel of the input feature layer during CNN operation. CBAM [27] proposes a simple and effective feedforward convolutional neural network attention module that considers two relatively independent dimensions of channel and space at the same time. Reference [28] proposes a new mobile network attention mechanism, which embeds location information into channel attention, avoiding excessive computational overhead, and can capture remote dependent information and save accurate location information.

3   Methods

Figure 1 shows the overall architecture diagram of our model using the Yolov7s model. We used ODConv and attention mechanisms in Backbone and Neck. Usually, introducing attention mechanisms adds an attention module, meaning that the model adds a layer and makes the model more complex. To avoid this problem, based on not breaking the architecture of the original Yolov7s as much as possible, we follow the way ODConv uses the attention mechanism. If we want to add an attention mechanism to a component, we can embed it directly into the component.

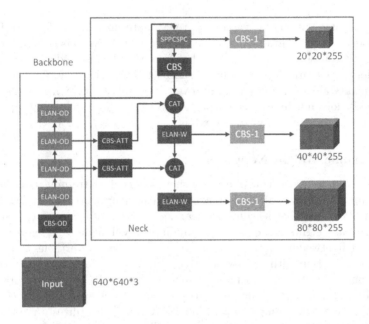

Fig. 1. Overall architecture diagram of YOLOv7 network. The model is mainly composed of three parts: Input, Backbone, and Head. This framework diagram only shows the parts we have modified, where - OD represents the use of ODConv and - ATT represents the use of attention mechanism. Since Yolov7 is all stacked by convolution, the output part has not been modified, so it is represented by CBS-1.

3.1 Network Architecture

The YOLOv7 network model is mainly composed of three parts: Input, Backbone, and Head. First, the image is input to the input layer. After a series of operations, such as slicing and data enhancement for pre-processing, it is sent to the backbone network to extract the corresponding features. Subsequently, the extracted features are fused in the Neck module at different scales. The fused features are fed to the inspection head, and the results are output after testing.

In the Backbone module, the regular convolution is replaced with ODConv to extract richer semantic features. ODConv can be defined as:

$$y = (\alpha_{w1} \odot \alpha_{f1} \odot \alpha_{c1} \odot \alpha_{s1} \odot W_1 + ... + \alpha_{wn} \odot \alpha_{fn} \odot \alpha_{cn} \odot \alpha_{sn} \odot W_n) \times x \quad (1)$$

where $\alpha_{wi} \in \mathbb{R}$ represents the attention scalar of the convolution kernel W_i ; $\alpha_{si} \in \mathbb{R}^{k \times k}$, $\alpha_{ci} \in \mathbb{R}^{c_{in}}$, $\alpha_{fi} \in \mathbb{R}^{c_{out}}$ represents the scalar attention calculated along the spatial dimension, input channel dimension and output channel dimension of the convolution kernel W_i, respectively. $\odot$ represents multiplication operations in different dimensions along kernel space.

We used the SENet attention module but calculated them with multiple heads as $\pi_i(x)$, the structure of which is shown in Fig. 2. Specifically, the input x is first compressed into a feature vector with length by channel-level global average pooling (GAP) operation. The FC layer maps compressed eigenvectors to a low-dimensional space with a reduction ratio of r. For the four head branches, each branch has an FC layer with output sizes $k \times k$, $c_{in} \times 1$, $c_{out} \times 1$, $n \times 1$, and a Softmax or Sigmoid function that generates normalized attention α_{si}, α_{ci}, α_{fi}, α_{wi}, respectively. In ODConv, for the convolution kernel W_i: (1) α_{si} assigns an attention scalar to each filter's convolution parameter within the spatial position; (2) α_{ci} assigns different attention scalars to the channel of each convolutional filter; (3) α_{fi} assigns different attention scalars to the convolution filter; (4) α_{wi} assigns an attention scalar to the entire convolution kernel.

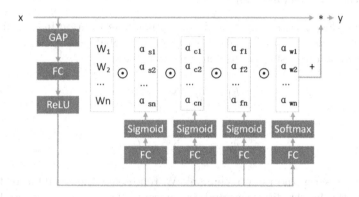

Fig. 2. ODConv uses a new multidimensional attention mechanism to compute four types of attention α_{si}, α_{ci}, α_{fi} and α_{wi} in the kernel space in parallel.

In principle, these four types of attention are complementary and gradually multiply them by convolutional kernels in the order of position, channel, filter, and core, capturing rich contextual information to provide performance guarantees.

3.2   Loss Function Optimization

We use the traditional Yolo loss: box regression loss, classification loss, and object loss. BBR loss is an important factor in whether the localization is accurate. The definition of IoU is crucial for calculating BBR losses. We are using WIoUv1, which is calculated as follows:

$$\mathcal{L}_{WIoUv1} = \mathcal{R}_{WIoU} \mathcal{L}_{IoU} \tag{2}$$

$$\mathcal{R}_{WIoU} = exp(\frac{(x - x_{gt})^2 + (y - y_{gt})^2}{(W_g^2 + H_g^2)^*}) \tag{3}$$

where,

$$\mathcal{L}_{IoU} = 1 - IoU = 1 - \frac{W_i + H_i}{S_u} \tag{4}$$

W_g and H_g are the size of the smallest closed box.

It can be seen from the formula that it has two layers of attention mechanism: (1)$\mathcal{R}_{WIoU} \in [1, e]$, which will significantly amplify the $\mathcal{L}_{IoU}$ of the ordinary mass anchor box;(2) $\mathcal{L}_{IoU} \in [0, 1]$, which will significantly reduce the $\mathcal{R}_{WIoU}$ of high-quality anchor frames, and focus on the distance between the center point when the anchor frame and the target frame coincide, which weakens the punishment of geometric factors to a certain extent, reduces training intervention, and improves the generalization ability of the model accordingly.

3.3   Attention Mechanism

Squeeze-and-Excitation (SE) attention does not take location information into account when using global pooling, so here the global pooling is decomposed into one-to-one dimension feature encoding operations:

$$z_c^h(h) = \frac{1}{W} \sum_{0 \leq i < W} x_c(h, i) \tag{5}$$

$$z_c^w(w) = \frac{1}{H} \sum_{0 \leq j < H} x_c(j, w) \tag{6}$$

where c represents the c-layer channel, W represents the weight direction, H represents the height direction, and z represents the output. This results in a pair of feature maps with directional awareness.

In order to obtain a feature map that is both channel-aware and sensitive to position information, the feature maps obtained in Eqs. 5 and 6 are first spliced together and then sent to a shared 1×1 convolution transformation function, the formula is:

$$f = \delta(F_1([z^h, z^w])) \tag{7}$$

where $[,]$ is series operations along spatial dimensions, the δ is the nonlinear activation function, and $f \subseteq \mathbb{R}^{\frac{C}{r} \times (H+W)}$ is the intermediate feature map that encodes spatial information in the horizontal and vertical directions. Then divide f into two independent tensors along the spatial dimension, and use two 1×1 convolution transformations F_h and F_w to transform the number of channels of f^h and f^w to be the same as the input X, respectively, with the formula:

$$g^h = \sigma(F_h(f^h)) \tag{8}$$

$$g^w = \sigma(F_w(f^w)) \tag{9}$$

where σ is the sigmoid function. The output g^h and g^w are expanded and used as attention weights, respectively. Finally, the output of our coordinate note block Y can be written as:

$$y_c(i, j) = x_c(i, j) \times g_c^h(i) \times g_c^w(j) \tag{10}$$

4   Experiments

In this section, we conduct corresponding experiments to evaluate the performance of our model. We also conducted ablation studies on the components to fully validate their effectiveness on our model.

4.1   Datasets and Implementation Details

To train our model, we used a dataset with a total of 3346 images, where the training set contains 2026 images, the validation set contains 630 images, and the test set contains 690 images. Before training, nine anchor box sizes suitable for our dataset have been obtained by the K-means clustering algorithm: 9×2, 16×4, 24×7, 39×13, 71×26, 104×51, 160×93, 238×151, 478×204.

We built our network on PyTorch and trained 500 epochs on a PC using two NVIDIA GeForce GTX 1080 Ti GPUs and 11 GB of memory.

4.2   Comparative Experiments

In order to verify the universality of our model and compare it with the latest model of the Yolo series, since the PPYOLO series only trained 36 epochs to get good results, we also only trained 36 epochs, and the results are shown in Table 1. Among them, the IoU threshold taken by AP^{test} is 0.75.

Table 1. Comparison of different models on the Coco dataset

Models	AP^{val}_{50-95}	AP^{val}	AP^{test}
Yolov7	/	51.2	51.4
PPYOLOE-S [29]	/	42.7	43.1
PPYOLOE-M [29]	/	48.6	48.9
PPYOLOE-L [29]	/	50.9	51.4
PPYOLOE-X [29]	/	51.9	52.2
Yolov8s [30]	44.9	/	/
ours	46.9	66.1	65.3

As can be seen from the above table, our model has improved AP by 14.9% compared to the original Yolov7, and the effect is very significant. At the same time, we also compared the latest Yolov8 model, which is 2.0% better. This shows that our model has a significant improvement in object detection in the Yolo field.

4.3   Ablation Experiments

To fully validate the effectiveness of each component in our model, we conducted an ablation study. W indicates that WIoU is used, O means that ODConv is used, S means that the SENet attention mechanism is introduced, C means that the CBAM attention mechanism is introduced, and CA means that the coordinate attention mechanism is introduced. These models were trained and tested with the same dataset and metrics, and the experimental results are shown in Table 2.

Table 2. Ablation experiments

Models	Precision	Recall	AP^{val}	AP^{test}
Yolov7	79.22	68.22	71.02	70.3
Yolov7s+W	81.11	68.41	71.17	71.3
Yolov7s+O	80.09	68.25	71.13	71.0
Yolov7s+S	79.47	68.07	70.03	68.3
Yolov7s+C	80.97	68.66	71.38	69.9
Yolov7s+CA	81.82	69.26	72.32	71.8
Yolov7s+W+O	82.11	68.92	71.89	72.1
Yolov7s+W+O+CA(ours)	83.07	69.85	72.39	72.3

As can be seen from the above table: (1) Each component we used has improved in performance, and the most significant improvement effect is the addition of an attention mechanism, followed by the use of WIoU and then the use of ODConv. This may be because the ODConv improvement effect on small objects is more significant, and only small potholes in the corners of some images are reflected in our dataset. The overall number may not be very large, so the improvement is not apparent. (2) In order to reflect the effectiveness of the attention mechanism we use, we use the SENet mechanism and the CBAM mechanism to compare, and it can be seen that the effect is improved one by one, and it is also verified that the use of four different types of attention can significantly improve performance, especially the accuracy, compared to the coordinate attention mechanism we use with SENet, which improves by 2.35%, which shows that our model can detect more accurate positive samples. (3) Continuous superposition of improving components, the performance also improves, showing that our improvement points do not conflict, and can jointly improve performance.

4.4   Visualize Results

We conducted qualitative experiments on the road depression dataset, compared our network with the original Yolov7 and Yolov8, and obtained the following visualizations. It can be seen that our model is superior to the other two models in terms of both detection and positioning accuracy (Fig. 3).

(a) Yolov7   (b) Yolov8   (c) ours

Fig. 3. Visualize Results

5 Conclusion

We propose a road detection algorithm based on Yolov7. Based on Yolov7, to make the positioning more accurate, we used WIoU, the total loss decreased, and the AP was improved. There will be more or less some small objects in the data set, so ODConv is introduced to improve the detection accuracy of this part. In the past, the attention mechanism was directly added to an attention component, which would make the model more complex, so we directly applied the attention mechanism to the existing component, which would not increase the complexity of the model and achieve the purpose of improving accuracy. In addition, we have conducted a large number of experiments to verify the superiority of our model.

Acknowledgment. This work was supported in part by the Natural Science Foundation of Hunan Province under grant 2021JJ30138.

References

1. Miller, J.S., Bellinger, W.Y.: Distress identification manual for the long-term pavement performance program (fifth revised edition). calibration (2014)
2. Fan, R., Ozgunalp, U., Wang, Y., Liu, M., Pitas, I.: Rethinking road surface 3-d reconstruction and pothole detection: from perspective transformation to disparity map segmentation. IEEE Trans. Cybern. **PP**(99) (2021)

3. Koch, C., Brilakis, I.: Pothole detection in asphalt pavement images. Adv. Eng. Inform. **25**(3), 507–515 (2011)
4. Chang, K.T., Chang, J.R., Liu, J.K.: Detection of pavement distresses using 3D laser scanning technology. In: International Conference on Computing in Civil Engineering (2005)
5. Lin, J., Liu, Y.: Potholes detection based on SVM in the pavement distress image (2010)
6. Liu, W., et al.: SSD: single shot multibox detector. In: Leibe, B., Matas, J., Sebe, N., Welling, M. (eds.) ECCV 2016. LNCS, vol. 9905, pp. 21–37. Springer, Cham (2016). https://doi.org/10.1007/978-3-319-46448-0_2
7. Zhu, L., Wang, X., Ke, Z., Zhang, W., Lau, R.W.H.: BiFormer: vision transformer with Bi-level routing attention. ArXiv abs/2303.08810 (2023)
8. Benjumea, A., Teeti, I., Cuzzolin, F., Bradley, A.: YOLO-Z: improving small object detection in YOLOv5 for autonomous vehicles (2021)
9. He, K., Zhang, X., Ren, S., Sun, J.: Deep residual learning for image recognition. IEEE (2016)
10. Huang, G., Liu, Z., Laurens, V.D.M., Weinberger, K.Q.: Densely connected convolutional networks. IEEE Computer Society (2016)
11. Jocher, G.: YOLOv5 by Ultralytics. https://doi.org/10.5281/zenodo.3908559, https://github.com/ultralytics/yolov5
12. Girshick, R., Donahue, J., Darrell, T., Malik, J.: Rich feature hierarchies for accurate object detection and semantic segmentation. IEEE Computer Society (2014)
13. Redmon, J., Divvala, S., Girshick, R., Farhadi, A.: You only look once: unified, real-time object detection. In: Computer Vision Pattern Recognition (2016)
14. Wang, C.Y., Bochkovskiy, A., Liao, H.Y.M.: YOLOv7: trainable bag-of-freebies sets new state-of-the-art for real-time object detectors. arXiv e-prints (2022)
15. Ma, N., Zhang, X., Zheng, H., Sun, J.: ShuffleNet V2: practical guidelines for efficient CNN architecture design. ArXiv abs/1807.11164 (2018)
16. Dollár, P., Singh, M., Girshick, R.: Fast and accurate model scaling (2021)
17. Lee, Y., Hwang, J.W., Lee, S., Bae, Y., Park, J.: An energy and GPU-computation efficient backbone network for real-time object detection. In: 2019 IEEE/CVF Conference on Computer Vision and Pattern Recognition Workshops (CVPRW) (2019)
18. Wang, C.Y., Bochkovskiy, A., Liao, H.Y.M.: Scaled-YOLOv4: scaling cross stage partial network. In: Computer Vision and Pattern Recognition (2021)
19. anonymous: Designing network design strategies (2022)
20. Zhang, Y.F., Ren, W., Zhang, Z., Jia, Z., Wang, L., Tan, T.: Focal and efficient IoU loss for accurate bounding box regression (2021)
21. Tong, Z., Chen, Y., Xu, Z., Yu, R.: Wise-IoU: bounding box regression loss with dynamic focusing mechanism. ArXiv abs/2301.10051 (2023)
22. Yang, B., Bender, G., Ngiam, J., Le, Q.V.: CondConv: conditionally parameterized convolutions for efficient inference (2019)
23. Chen, Y., Dai, X., Liu, M., Chen, D., Liu, Z.: Dynamic convolution: attention over convolution kernels. In: 2020 IEEE/CVF Conference on Computer Vision and Pattern Recognition (CVPR) (2020)
24. Li, C., Zhou, A., Yao, A.: Omni-dimensional dynamic convolution. ArXiv abs/2209.07947 (2022)
25. Hu, J., Shen, L., Sun, G., Albanie, S.: Squeeze-and-excitation networks. In: IEEE (2017)
26. Mnih, V., Heess, N., Graves, A., Kavukcuoglu, K.: Recurrent models of visual attention. Adv. Neural Inf. Process. Syst. **3** (2014)

27. Woo, S., Park, J., Lee, J.-Y., Kweon, I.S.: CBAM: convolutional block attention module. In: Ferrari, V., Hebert, M., Sminchisescu, C., Weiss, Y. (eds.) ECCV 2018. LNCS, vol. 11211, pp. 3–19. Springer, Cham (2018). https://doi.org/10.1007/978-3-030-01234-2_1
28. Hou, Q., Zhou, D., Feng, J.: Coordinate attention for efficient mobile network design (2021)
29. Xu, S., et al.: PP-YOLOE: an evolved version of YOLO (2022)
30. Jocher, G., Chaurasia, A., Qiu, J.: YOLO by Ultralytics. https://github.com/ultralytics/ultralytics

Explicit Exploring Geometric Modality for Shape-Enhanced Single-View 3D Face Reconstruction

Shikun Zhang, Haoran Xu, Fengyi Song$^{(\boxtimes)}$, Ge Song, and Ming Yang

School of Computer and Electronic Information/School of Artificial Intelligence, Nanjing Normal University, Nanjing, Jiangsu, China
{zsk.work,hrxu,f.song,g.song,myang}@njnu.edu.cn

Abstract. For 3D shape inference, single-view 3D Face Reconstruction heavily relies on capturing geometric structure information from the 2D face. Most 3DMM-based approaches learn the geometry parameters directly from the 2D image appearance, but the limited information makes it an ill-conditioned task and makes the model struggle to learn the inference evidences. In this work, we propose that the 2D face boundary image contains more semantic information in the face contour connecting lines and can represent the basic geometric structure of the face. In addition, we propose an Explicit Geometric Modality Network (EGMNet) for enhancing face shape inference. The EGMNet consists of the *appearance induced shape branch* and the *boundary induced shape branch*. The former is built with an FPN-based network capable of combining multi-scale features, while the latter is built with a light-weighted network capable of extracting rich shape features. Finally, these two modalities' induced shape features are combined to achieve shape-enhanced 3D face reconstruction.

Keywords: 3D Face Reconstruction · Geometric Modality · Shape enhancement · 3D Morphable Model

1 Introduction

3D face reconstruction is an important task in the fields of computer vision and image processing. Currently, 3D face reconstruction contributes greatly to many visual analysis tasks, including medical analysis, virtual reality, and face-concentrated tasks such as face key point detection [21], face recognition [2], and face editing [15]. Most existing works follow the famous framework of the 3D Morphable Model (3DMM), which provides an efficient and effective solution for 3D face reconstruction.

Deep learning has facilitated the development of many advanced learning tasks, and it also shows great potential for 3D face reconstruction tasks in the 3DMM framework. However, although the emergence of deep learning has

E. Chen et al. (Eds.): BigData 2023, CCIS 2005, pp. 88–98, 2023.
https://doi.org/10.1007/978-981-99-8979-9_7

enhanced the learning capability of facial features, the amount of effective information about faces contained in a single 2D face image is always limited, which is a challenge that cannot be ignored for 3D face reconstruction tasks. Especially when the target face presents a large pose or occlusion, the incomplete face information has a more serious impact on the related tasks. Therefore, in order to fully exploit the face geometry information, Shang et al. [14] proposed a self-supervised training architecture that uses the geometric consistency of multiple views to constrain the pose and depth estimation of faces, thus realizing geometrically accurate face reconstruction. Li et al. proposed a multi-attribute regression network (MARN) [10], the work design the geometric contour constraint loss function, using the constraints of sparse 2D face landmarks to improve the reconstructed geometric contour information.

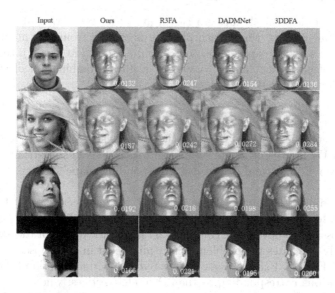

Fig. 1. 3D reconstruction results of different methods [7,8,20] for comparison, the reconstructed NME results are shown in the lower right corner. The lower the better.

In fact, in the field of super-resolution image reconstruction [1,5,19], many scholars have enhanced the localization of edge points by adding edge images of the original image to the image reconstruction process. Because compared with the original 2D pixel images, the face contour edge images not only contain facial landmark points, but also contain more semantic information in the connecting lines, which can provide a richer facial geometric structure. Inspired by the above work and the need to fully exploit the facial feature information, it is not difficult to think that the facial boundary represents the basic contour of the face and even has a large space to characterize the facial pose, which is helpful for the localization and pose regression of key facial points.

In this work, to improve the accuracy of facial shape and pose, we propose to incorporate facial edge images into the 3D face reconstruction task to achieve enhancement of edge features. Meanwhile, we propose a two-branch face reconstruction network based on shape enhancement, which is used to enhance the ability of the model to extract geometric details. Considering the efficiency of the model and the ability to extract features, we chose the lighter mobilenet module and the multi-scale feature fusion module (MFFM) to construct the two branches of the network separately, and finally combined the features extracted from the two branches to infer the final 3D facial structure.

2   Method

2.1   Preliminary: 3DMM and Projection

3DMM [2] is a classical 3D face statistical model for recovering 3D face shapes from 2D facial images, where each 3D face is viewed from two main kinds of components, i.e., shape and expression. Then, as shown in Eq. (1), the 3D representation of each face can be viewed as a linear combination of shape and expression, denoted as W_s and W_e.

$$S = \bar{S} + \alpha_s W_s + \alpha_e W_e \tag{1}$$

where $\bar{S}$ denotes the mean 3D face, and the shape basis vector W_s and expression basis vector W_e are the principal component bases derived statistically from real 3D facial scan data. α_s and α_e are the corresponding shape and expression basis coefficients.

After obtaining the rebuilt 3D face shape S, the perspective projection model [21] is used to project the 3D surface point S onto the 2D image plane. This makes it possible to obtain the complete 2D landmark points of the target face, which largely alleviates the problem of face alignment in large poses. For details, as shown in Eq. (2), V_{2d} is a weak perspective projection function for obtaining the 2D positions corresponding to the 3D model vertices, where f is a scaling factor, Pr is a fixed orthogonal projection matrix, and R and t_{2d} denote the rotation matrix and translation vector, respectively.

$$V_{2d} = Pr * f * R * S + t_{2d} \tag{2}$$

Thus, the network based on the 3DMM model, which finally requires solving the 62-dimensional parameters $P = [f, R, t_{2d}, \alpha_s, \alpha_e]$, the pose parameters $\{f, R, t_{2d}\}$ are a total of 12-dimensional parameters, α_s is a shape factor of 40 dimensions, and α_e is an expression factor of 10 dimensions.

3   Network

We propose a boundary mode-enhanced shape inference model, named Explicit Geometric Modal Network (EGMNet). The network regresses the 3DMM reconstruction parameters from two different concerns, namely global features and

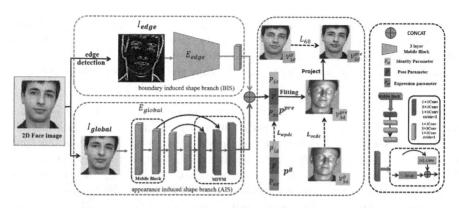

Fig. 2. The architecture of the proposed EGMNet model, contains two main branches, including the appearance induced shape branch (AIS) and the boundary induced shape branch (BIS). Both branches are designed based on the lighter mobile blocks in MobileNet-V2 [13] for efficiency, while BIS is used to extract shape-specific features. To obtain richer information features, we designed a multi-scale feature fusion module (MFFM) in the AIS branch. The loss criteria used and the detailed network structure are shown on the right side of the figure.

facial geometric boundary features, using different representations of 2D face images. The proposed EGMNet, as shown in Fig. 2, is made up of two main feature extraction branches: the appearance induced shape branch (AIS) and the boundary induced shape branch (BIS). First, the 2D face image input to the network is detected by the edge detection task [4] to obtain the I_{edge}, the algorithm uses the difference between two adjacent pixels in the diagonal direction to approximate the gradient magnitude to detect edges, and detects vertical edges better than edges in other directions, with higher localisation accuracy. For the input 2D face image $f(x, y)$, the output image is $g(x, y)$ after the edge detection process. As shown in Eq. (3)

$$g_x = f(x, y) - f(x + 1, y + 1)$$

$$g_y = f(x + 1, y) - f(x, y + 1)$$

$$g(x, y) = \sqrt{g_x^2 + g_y^2}$$

(3)

where g_x and g_y denote the gradient of the input point (x, y) corresponding to the x-axis and y-axis, respectively, and $g(x, y)$ is calculated as the joint gradient of that point. Compared to the original input image, I_{edge} has a more pronounced edge geometry structure, which provides more direct shape and pose information. Therefore, I_{edge} is used as a new modal image input into the network for learning.

Then, AIS and BIS branches are used to extract features from the global (I_{global}) and boundary 2D face images (I_{edge}), respectively. BIS is mainly used to extract the boundary geometry information from I_{edge} for feature enhancement.

$$p = F\{CONCAT[E_{edge}(I_{edge}), E_{global}(I_{global})]\}$$

(4)

In Eq. (4), E_{edge} and E_{global} are the feature extraction modules in BIS and AIS branches. Finally, the extracted information is combined and used to reason about the final 3DMM reconstruction parameters p^{pre}. p^{pre} is the 62-dimensional reconstruction parameter $[f, R, t_{2d}, \alpha_s, \alpha_e]$ that the model needs to regress, as mentioned in Sect. 2.

Network Structure. To make the network model lighter and more efficient, we built the E_{edge} and E_{global} branches using the mobile blocks in MobileNet-V2 [13]. As shown in the right part of the Fig. 2, a mobile block contains four computational modules. The depth-separated convolution used by the mobile block, which consists primarily of 1×1 pointwise convolution, 3×3 depthwise convolution, and 1×1 pointwise convolution. This design allows the model to use a smaller number of parameters and to compute more efficiently.

From the right part of Fig. 2, the E_{edge} branch is used to process the input face boundary geometry image I_{edge}. Since I_{edge} does not carry much depth information, we design E_{edge} as a shallow network consisting of three basic layer blocks, each layer consists of a mobile block. So E_{edge} is mainly used to extract more potential features containing facial geometry and pose details from I_{edge}. Considering that the original 2D image contains more information, in order to balance learning efficiency and learning depth at the same time, we use four mobile blocks to build the E_{global} branch, and we specifically build a multi-scale feature fusion module (MFFM). The MFFM module takes reference from the pyramid network [11] for multi-scale feature fusion to obtain richer information by combining multi-layer features. The main structure of MFFM is shown in the lower right corner of Fig. 2. Firstly, the input features of the previous layer are up-sampled using two-fold inverse convolution, and then the corresponding low-level features are altered using 1×1 convolution to have the same number of channels as the up-sampled features, and then finally the corresponding elements of the two features are summed to complete the feature fusion. The corresponding upper layer fused feature is used as the next layer fused high-level feature, until finally the final multi-stage fusion feature is obtained. Finally, we combine the information extracted from the two branches to achieve a more accurate information representation. Specifically, the multi-stage fused features of the MFFM are summed with the features obtained from the facial boundary contours (E_{edge}). The fused features are input into a fully connected layer, which inference the 3DMM reconstruction parameters $[f, R, t_{2d}, \alpha_s, \alpha_e]$ and use this 3DMM reconstruction parameters to regress the 3D shape of the target face.

4   Loss Criteria

We first use the weighted parametric distance cost (WPDC) [21] for constraining the model in order to reason about more accurate 3DMM reconstruction parameters. As shown in Eq. (5) where p^g is the ground truth 3D parameter and p is the 3D parameter predicted by the model. W is a matrix of weight variables indicating the importance of each dimensional parameter in p.

$$L_{wpdc} = (p^g - p)^T W (p^g - p)$$
$$W = diag(w_1, w_2, ..., w_{62}) \tag{5}$$
$$w_i = \frac{\|V_{2d}(\hat{p}_i) - V_{2d}(p^g)\|}{\max(W)}$$

Then we intend to take the depth into consideration. Chamfer Distance [6] as Eq. (6) is introduced to calculate between sets of 3D vertices.

$$L_{vcdc} = \frac{1}{N} \sum_{x \in V_{3d}(p^g)} \min_{y \in V_{3d}(p)} \|x - y\|^2 + \tag{6}$$

$$\frac{1}{N} \sum_{x \in V_{3d}(p)} \min_{y \in V_{3d}(p^g)} \|x - y\|^2$$

Here, N means the number of vertices and V_{3d} could be calculated by Eq. (2) without Pr. During calculation, for each vertex in one set, find the minimum distance in the other set and sum the square distances up as this type of loss.

To enhance the representation of face geometric information, we use sparse 2D face landmarks as a weak constraint to further constrain the basic contours of the face as follows.

$$L_{68} = \|(f * R * Pr * S_{68}(\alpha_{id}, \alpha_{ex}) + t_{2d}) - \tag{7}$$
$$(f^g * R^g * Pr * S_{68}(\alpha_{id}^g, \alpha_{ex}^g) + t_{2d}^g)\|^2$$

The parameters in Eq. (7) and the parameters in Eq. (2) have the same representation, α_{id}^g, and α_{ex}^g, f^g, R^g and t_{2d}^g are the ground truth values of the corresponding parameters.

The final loss of our network is shown in Eq. (8). Where λ_{wpdc}, λ_{vcdc} and λ_{68} are used to balance the weights of these constraints.

$$L_{total} = \lambda_{wpdc} L_{wpdc} + \lambda_{vcdc} L_{vcdc} + \lambda_{68} L_{68} \tag{8}$$

5 Experiments

In this section, we focus on the training details of our network on the 300W-LP [20] dataset, and then evaluate the performance of our proposed method on dense face alignment and 3D face reconstruction tasks, and compare it with the latest methods on the AFLW2000-3D and AFLW test datasets.

5.1 Training Details

We use the PyTorch deep learning framework to train the model and use SGD to optimize our network with a momentum of 0.9 and a weight decay of 5e-4. We empirically set the initial learning rate to 0.01 and specified a batch size of 64. Our loss function weights λ_{wpdc} is set to 1, λ_{68} and λ_{vcdc} are set to 3.

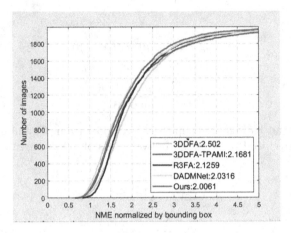

Fig. 3. Cumulative errors distribution (CED) curves for 3D face reconstruction on AFLW2000-3D with [7,8,20,21]. The mean NME (%) of each method is shown in the bottom legend.

5.2   3D Face Reconstruction

In this section, we compare the performance of 3D face reconstruction on the AFLW2000-3D dataset with methods [7,8,20,21], which have the same pipeline and final vertex count as ours. The NME metrics are computed between the aligned 3D geometries for performance evaluation. As shown in Fig. 3, the 3D shape reconstructed with our method outperforms several of the above methods. As shown in Fig. 1, our reconstruction results are smoother and have more details. Note that for uncertain segmentation of the input image, the reconstruction of [10] is random, so we did not make a fair comparison with this method in this experiment.

5.3   3D Face Alignment Results

Similarly, to evaluate the performance of the model on the face-dense alignment task, we used the normalized mean error (NME) as an evaluation metric, tested on the AFLW and AFLW2000-3D datasets. In the test, we divided the AFLW and AFLW2000-3D datasets into three gradients according to yaw angle, namely $[0°, 30°], [30°, 60°]$ and $[60°, 90°]$, and compared the performance of the models on these three gradients. The results of the comparison on the AFLW dataset are shown in Table 2, and the results of the comparison on the AFLW2000-3D dataset are shown in Table 1, and please note that these values are recorded from their published paper and highlight the best results. Compared to most of the facial alignment methods, our method is competitive in terms of average, or different yaw angles. As shown in Fig. 4, we show the visualisation results of the method in this paper and other methods [7,8] for the face alignment task on the AFLW2000-3D dataset. We have selected as input images the target face in front face, side face and different region occlusion states, as shown in the first column of Fig. 4. Each method displays two outputs, the first one is the reconstruction result visualisation image and the second one is the keypoint alignment visualisation result. The comparison reveals that the method proposed in this paper not only achieves better reconstruction results, but also possesses a smaller error in the sparse point alignment task. Especially in the performance of specific details, such as the key point alignment of the chin contour and the key point alignment of the mouth part.

Table 1. The NME (%) of 2D face alignment for different range of poses on AFLW2000-3D with the first good results are highlighted, the lower is better.

Method	AFLW2000-3D Dataset (68 pts)			
	$[0°, 30°]$	$[30°, 60°]$	$[60°, 90°]$	Mean
RCPR [3]	4.260	5.960	13.180	7.800
SDM [17]	3.670	4.940	9.670	6.120
3DDFA (CVPR16) [21]	3.780	4.540	7.930	5.420
Yu et al. [18]	3.620	6.060	9.560	6.410
DAMDNet (ICCVW19) [7]	2.907	3.830	4.953	3.897
MARN (ICPR21) [10]	2.989	3.670	4.613	3.757
MFIRRN (ICASSP21) [9]	2.841	3.572	**4.561**	3.658
EOSNet (ICIP2022) [16]	2.906	3.739	4.677	3.774
Ours	**2.692**	**3.444**	4.656	**3.598**

Table 2. The NME (%) of 2D face alignment for different range of poses on AFLW with the first good results are highlighted, the lower is better.

Method	AFLW Dataset (21 pts)			
	$[0°, 30°]$	$[30°, 60°]$	$[60°, 90°]$	Mean
RCPR [3]	5.430	6.580	11.530	7.850
SDM [17]	4.750	5.550	9.340	6.550
DEFA [12]	4.500	5.560	7.330	5.803
3DDFA (CVPR16) [21]	5.000	5.060	6.740	5.600
DAMDNet (ICCVW19) [7]	4.359	5.209	6.028	5.199
MARN (ICPR21) [10]	4.306	4.965	5.775	5.015
MFIRRN (ICASSP21) [9]	4.321	5.051	5.958	5.110
EOSNet (ICIP2022) [16]	4.212	4.935	5.787	4.978
Ours	**4.132**	**4.909**	**5.775**	**4.939**

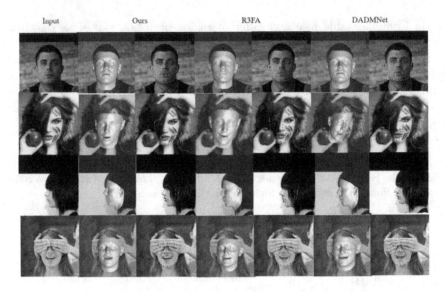

Fig. 4. 2D face alignment results of different methods [7,8] for comparison.

Table 3. Ablation study for validating the efficiency of the boundary induced shape branch (BIS) and multi-scale feature fusion module (MFFM).

Network	AFLW (mean)	AFLW2000-3D (mean)
EGMNet (w/o MFFM, BIS)	5.228	4.008
EGMNet (w/o BIS)	5.157	3.851
EGMNet (w/o MFFM)	5.064	3.894
EGMNet	**4.939**	**3.598**

5.4 Ablation Study

In this section, we test the degree of impact of each module on the model's performance, focusing on the impact of the multi-scale feature fusion module (MFFM) in the AIS branch, and the boundary induced shape branch (BIS). As shown in Table 3, alignment results on both datasets are used as a reference. In addition, in order to verify the usefulness of the three loss functions in training, we add the three loss functions one by one in training and test the accuracy of the model on the face alignment task separately. As shown in Table 4, each part trains the same epoch, WPDC is mainly used to constrain the model to regress out the accurate 3DMM reconstruction parameters, it is used in the first step of model training. VCDC is constrained in 3D vertex space to improve the accuracy of model alignment in 3D vertex space. Finally, the 2D sparse point distance loss L_{68} is used for weak supervision.

Table 4. Ablation experiments to verify the role of three loss functions in training, shows the results of face alignment on both datasets

Loss Criteria	AFLW (mean)	AFLW2000-3D (mean)
WPDC	6.602	5.239
WPDC+VCDC	5.060	3.876
WPDC+VCDC+L_{68}	**4.939**	**3.598**

6 Conclusion

In this paper, we introduce face geometric boundary images as new representational features for the single-view 3D face reconstruction task. Compared with the traditional model, our model improves the accuracy of pose parameters and shape parameters for model regression by fully exploiting the contour boundary features. A new approach for 3D face reconstruction is also proposed in the network design. Experiments conducted on the corresponding dataset demonstrate the effectiveness of our proposed method.

Acknowledgements. This work is supported in part by China Postdoctoral Science Foundation (No. 2017M621749)' National Natural Science Foundation of China (No. 62106108, 62076135, 61876087, 62276138)' Natural Science Foundation of Jiangsu Province (No. BK20210559)' and Natural Science Research of Jiangsu Higher Education Institutions (No. 21KJB520012).

References

1. Ai, W., Tu, X., Cheng, S., Xie, M.: Single image super-resolution via residual neuron attention networks. In: 2020 IEEE International Conference on Image Processing (ICIP), pp. 1586–1590 (2020). https://doi.org/10.1109/ICIP40778.2020.9191314
2. Blanz, V., Vetter, T.: A morphable model for the synthesis of 3D faces. In: Proceedings of the 26th Annual Conference on Computer Graphics and Interactive Techniques, pp. 187–194 (1999)
3. Burgos-Artizzu, X.P., Perona, P., Dollár, P.: Robust face landmark estimation under occlusion. In: Proceedings of the IEEE International Conference on Computer Vision, pp. 1513–1520 (2013)
4. Canny, J.: A computational approach to edge detection. IEEE Trans. Pattern Anal. Mach. Intell. **6**, 679–698 (1986)
5. Dong, C., Loy, C.C., He, K., Tang, X.: Image super-resolution using deep convolutional networks. IEEE Trans. Pattern Anal. Mach. Intell. **38**(2), 295–307 (2016). https://doi.org/10.1109/TPAMI.2015.2439281
6. Fan, H., Su, H., Guibas, L.J.: A point set generation network for 3D object reconstruction from a single image. In: Proceedings of the IEEE Conference on Computer Vision and Pattern Recognition, pp. 605–613 (2017)
7. Jiang, L., Wu, X.J., Kittler, J.: Dual attention MobDenseNet (DAMDNet) for robust 3D face alignment. In: Proceedings of the IEEE/CVF International Conference on Computer Vision Workshops, pp. 504–513 (2019)

8. Jiang, L., Wu, X.-J., Kittler, J.: Robust 3D face alignment with efficient fully convolutional neural networks. In: Zhao, Y., Barnes, N., Chen, B., Westermann, R., Kong, X., Lin, C. (eds.) ICIG 2019. LNCS, vol. 11902, pp. 266–277. Springer, Cham (2019). https://doi.org/10.1007/978-3-030-34110-7_23

9. Li, L., Li, X., Wu, K., Lin, K., Wu, S.: Multi-granularity feature interaction and relation reasoning for 3D dense alignment and face reconstruction. In: IEEE International Conference on Acoustics, Speech and Signal Processing, pp. 4265–4269. IEEE (2021)

10. Li, X., Wu, S.: Multi-attribute regression network for face reconstruction. In: 2020 25th International Conference on Pattern Recognition (ICPR), pp. 7226–7233. IEEE (2021)

11. Lin, T.Y., Dollár, P., Girshick, R., He, K., Hariharan, B., Belongie, S.: Feature pyramid networks for object detection. In: 2017 IEEE Conference on Computer Vision and Pattern Recognition (CVPR), pp. 936–944 (2017). https://doi.org/10.1109/CVPR.2017.106

12. Liu, Y., Jourabloo, A., Ren, W., Liu, X.: Dense face alignment. In: Proceedings of the IEEE International Conference on Computer Vision Workshops, pp. 1619–1628 (2017)

13. Sandler, M., Howard, A., Zhu, M., Zhmoginov, A., Chen, L.C.: MobileNetV2: inverted residuals and linear bottlenecks. In: Proceedings of the IEEE Conference on Computer Vision and Pattern Recognition, pp. 4510–4520 (2018)

14. Shang, J., et al.: Self-supervised monocular 3D face reconstruction by occlusion-aware multi-view geometry consistency. In: Vedaldi, A., Bischof, H., Brox, T., Frahm, J.-M. (eds.) ECCV 2020. LNCS, vol. 12360, pp. 53–70. Springer, Cham (2020). https://doi.org/10.1007/978-3-030-58555-6_4

15. Thies, J., Zollhofer, M., Stamminger, M., Theobalt, C., Nießner, M.: Face2Face: real-time face capture and reenactment of RGB videos. In: Proceedings of the IEEE Conference on Computer Vision and Pattern Recognition, pp. 2387–2395 (2016)

16. Wang, J., Zhang, S., Song, F., Song, G., Yang, M.: Exploring occlusion-sensitive deep network for single-view 3D face reconstruction. In: 2022 IEEE International Conference on Image Processing (ICIP), pp. 1821–1825 (2022). https://doi.org/10.1109/ICIP46576.2022.9897209

17. Xiong, X., De la Torre, F.: Global supervised descent method. In: Proceedings of the IEEE Conference on Computer Vision and Pattern Recognition, pp. 2664–2673 (2015)

18. Yu, R., Saito, S., Li, H., Ceylan, D., Li, H.: Learning dense facial correspondences in unconstrained images. In: Proceedings of the IEEE International Conference on Computer Vision, pp. 4723–4732 (2017)

19. Zhang, Y., Wu, Y., Chen, L.: MSFSR: a multi-stage face super-resolution with accurate facial representation via enhanced facial boundaries. In: 2020 IEEE/CVF Conference on Computer Vision and Pattern Recognition Workshops (CVPRW), pp. 2120–2129 (2020). https://doi.org/10.1109/CVPRW50498.2020.00260

20. Zhu, X., Lei, Z., Liu, X., Shi, H., Li, S.Z.: Face alignment across large poses: a 3D solution. In: Proceedings of the IEEE Conference on Computer Vision and Pattern Recognition, pp. 146–155 (2016)

21. Zhu, X., Liu, X., Lei, Z., Li, S.Z.: Face alignment in full pose range: a 3D total solution. IEEE Trans. Pattern Anal. Mach. Intell. **41**(01), 78–92 (2019)

Fine Edge and Texture Prior Guided Super Resolution Reconstruction Network

Peng Sun⬦, Jialuo Xu⬦, Shuaishuai Dong⬦, and Yi Chen(✉)⬦

School of Computer and Electronic Information/School of Artificial Intelligence,
Nanjing Normal University, Nanjing 210023, China
`cs_chenyi@njnu.edu.cn`

Abstract. In recent years, significant progress has been made in the field of super-resolution through the use of neural networks. Prior knowledge, such as edges and textures, is commonly incorporated into super-resolution reconstruction networks. However, existing models rely on fixed operators to extract binary edge and texture information, which often capture only rough features and fail to accurately represent the desired edge and texture characteristics. Consequently, this may result in the generation of spurious edges and difficulties in reconstructing image texture details. In this study, we propose a novel super-resolution neural network composed of three branches, with two branches specifically dedicated to extracting fine edges and textures. These two branches take the edge map and texture map of the high-resolution image as the target image, respectively, and are able to construct end-to-end neural networks through loss function constraints. Experimental results demonstrate the superiority of our model in reconstructing sharper edges and finer textures on benchmark datasets, including Set5, Set14, BSDS100, Urban100.

Keywords: Image super-resolution · Prior knowledge · Edge · Texture

1 Introduction

Super Resolution (SR) reconstruction has been a popular research area for the past few decades. The super resolution techniques aim to reconstruct Low Resolution (LR) images to High Resolution (HR) images, and while improving the quality of our life. There are many applications based on super-resolution technology, such as video reconstruction [3,4], social security [22], medical image enhancement [14,16],and military remote sensing [26].

With the development of deep learning techniques, super-resolution networks have achieved rapid progress in the field of single image super-resolution (SISR) for the past few years. The SRCNN [33] network was the first neural network designed for this task and showed better results than traditional methods using only three convolutional layers. Researchers later developed deeper networks like VDSR [15], EDSR [18], and MDSR, which achieved even better reconstruction

© The Author(s), under exclusive license to Springer Nature Singapore Pte Ltd. 2023
E. Chen et al. (Eds.): BigData 2023, CCIS 2005, pp. 99–111, 2023.
https://doi.org/10.1007/978-981-99-8979-9_8

outcomes. These models mostly extract information directly from low-resolution images to generate high-resolution images, but LR contain limited pixel information which makes it difficult to reconstruct higher quality image information.

Some studies have shown that incorporating prior knowledge, such as edge priors [30,36] and texture priors [28], can provide additional pixel information and improve the reconstruction quality of images. For example, Yang et al. [32] proposed a method that combines edge maps with LR images for super-resolution (SR) reconstruction. Fang et al. [11] introduced an edge network to reconstruct image edges and learn edge features. Wang et al. [28] utilized fixed-class texture priors to effectively reconstruct the texture details in images. Although these methods have achieved good reconstruction results by leveraging prior knowledge, they often overlook the differences between textures and edges, as well as the repetitive nature of textures.

Therefore, we utilize the GLCM (Gray-Level Co-occurrence Matrix) to extract the most frequently occurring texture features in the image. Additionally, we employ an edge detection operator to extract fine-grained edge information. As shown in Fig. 1, the extracted edges and textures differ significantly from the binary edge operator. Although the extracted texture map contains sufficient edge information, there are some spurious edges, so we utilize a refined edge constraint reconstruction to fuse them with features. We use these extracted texture and edge maps as the target images for the edge and texture branches of our network. By incorporating loss constraints, we construct an end-to-end network. In summary, our contributions are as follows:

1. We proposed a novel super-resolution network consists of three branches designed to extract fine-grained edge and texture information. By fusing the extracted edge and texture information, not only the internal texture of the image can be reconstructed, but also the problem of false edges can be solved using the fine-grained edge map. Numerous experiments demonstrate that our network helps to guide the super-resolution reconstruction, thus effectively solving the difficulties of image edge blurring and internal texture reconstruction.
2. We have devised a novel loss function incorporating three components: image content, edge, and texture losses. This integrated loss structure guides our model to converge effectively, enabling accurate reconstruction of image edges and texture details.

2   Related Works

2.1   Single Image Super-Resolution

In recent years, super-resolution techniques have been widely used in the field of single image super-resolution. The development of single image super-resolution can be divided into the following two steps:

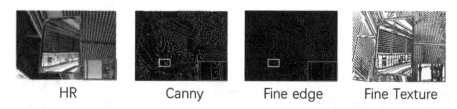

HR   Canny   Fine edge   Fine Texture

Fig. 1. The figure represents the edge map extracted by Canny, the refined edge map and the texture map used in this paper, respectively.

In the early stages of super-resolution research, conventional methods relied on techniques like linear interpolation and bicubic interpolation for image reconstruction. These interpolation algorithms leveraged neighboring pixel values to estimate missing pixels. While these approaches offered simplicity and flexibility, they faced challenges in accurately reconstructing high-frequency details in super-resolution images.

Subsequently, learning-based methodologies emerged to address the LR-to-HR mapping challenge. These approaches encompassed a range of techniques, including sparse-based methods [31], neighborhood embedding methods [7,27], random forests [23], and notably, convolutional neural networks (CNNs) [33]. With the rapid advancements witnessed in CNN research, they have risen to prominence, establishing themselves as the prevailing approach in the field. Prominent CNN models such as EDSR [18], RDN [35], SAN [8], and RFA [19] have garnered substantial attention, owing to their remarkable performance in image super-resolution tasks, thus solidifying their position at the forefront of the field.

2.2   Prior Information Assisted Image Reconstruction

In the past few years, super-resolution networks based on prior information have had a great impact on the field of super-resolution. Usually, a complex image contains many edge regions, so the introduction of an edge prior will have an important impact on the reconstruction of complex images. Tai et al. [25] proposed to combine the advantages of edge-directed SR and learning-based SR. Yang et al. [32] proposed an edge-guided recursive residual network (DEGREE) that introduces image edges into a neural network model. The network uses a bicubic interpolation preprocessed LR image as input and uses existing operators (e.g. Sobel detector [10], Canny detector [6] etc.), which introduced additional noise and generates artifacts. Sun et al. [24] used a novel gradient profile prior for super-resolution reconstruction. Li et al. [17] proposed to use edge information to introduce an encoder decoder to reconstruct high-resolution images. Fang et al. [11] proposed the soft-edge information extracted by Edge-Net, which solved the problem of fake edge appearance compared to the ready-to-use edge extractor. Zhao et al. [35] proposed IEGSR to accomplish super-resolution reconstruction using the high-frequency information of the image in the edge region.

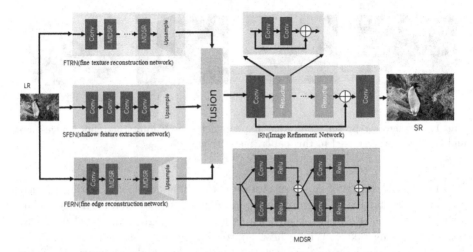

Fig. 2. The general composition of the FETSR network consists of four parts: shallow feature extraction network (SFEN), fine texture reconstruction network (FTRN), fine edge reconstruction network (FERN) and Image Refinement Network (IRN).

3   Methodology

3.1   Architecture

The progress of the reconstruction can be divided into the following two steps in our network. As seen in Fig. 2, first, we reconstruct the rough features by the SFEN, fine-edge extraction through the FERN and the texture information extracted by the FTRN. FERN and FTRN contain mainly the MDSR module, which is capable of fully extracting multi-scale edge and texture features through convolutional kernel-size-agnostic convolution. Second, we will concatenate and fuse the fine edge, rough features and fine texture information. The fused image tensor is then fed to the image refinement network and used to recover high quality images. In detail, FETSR consists of four modules: shallow feature extraction network (SFEN), a fine edge reconstruction network (FERN), a fine texture reconstruction network (FTRN), and an image refinement network (IRN).

In the first stage, the output of these network can be described as:

$$f_{rough} = F_{SF}(I_{LR}) \tag{1}$$

$$f_{edge} = F_{Edge}(I_{LR}) \tag{2}$$

$$f_{texture} = F_{texture}(I_{LR}) \tag{3}$$

where I_{LR} is the low-resolution image, F_{SF}, F_{edge} and $F_{texture}$ denote the SFEN, FERN and FTRN. f_{rough}, f_{edge} and $f_{texture}$ represent the shallow features, the image fine edge, and the image texture. Then they use fusion layers for merging:

$$f_{fusion} = F_{fusion}([f_{rough}, f_{edge}, f_{texture}]) \tag{4}$$

where [] operation represents the connection of the feature maps and $F_{fusion}()$ denotes the fusion layer, which can achieve features for fusion. In the second stage, we train the fused image tensor using the refinement network.

$$I_{SR} = F_{IRN}(f_{fusion}) \tag{5}$$

where I_{SR} is the reconstructed SR image and $F_{IRN}()$ represents the image refinement network. In training our network, we propose the following loss function to assist the reconstruction process.

$$L_{loss} = L_{content} + \lambda_1 L_{edge} + \lambda_2 L_{texture} \tag{6}$$

where λ_1 and λ_2 are hyper-parameters, $L_{content}$, L_{edge} and $L_{texture}$ denote the loss, and this will be discussed in the following chapters.

3.2   Shallow Feature Extraction Network (SFEN)

First, we use SFEN to extract rough image feature. As a general rule that the rough image features can be easily detected, so a convolutional layer using a 3×3 convolutional kernel is applied to map the image to a high dimension. Then, the low-frequency information of SR image is extracted by using five identical convolutional layers, each layer is represented as:

$$f_n = w_n * f_{n-1} + b_n(n = 1, 2, \cdots, 5) \tag{7}$$

where f_n, w_n and b_n represent the feature maps, weights and biases of the current convolutional layer output, respectively. f_{n-1} is the output of the upper layer and it will feed into the current layer, where n varies from 1 to 5. A related study found that sufficient shallow features can be extracted with 5 layers of convolution. Finally, when n = 5, we utilize an upsample module to upscale the extracted features to the same dimension of the HR.

$$f_{rough} = F_{up}(f_5) \tag{8}$$

where F_{up} denotes the up-sample module, which consist of a sub-pixel layer and employ two convolutional layers for the transition to the image refinement network. The output f_{rough} represents the image features extracted by convolution at a shallow level.

3.3   Fine Texture Reconstruction Network (FTRN)

A prior information is often used for image reconstruction and has led to significant improvements in image quality. The texture prior is introduced in FTRN, which can reconstruct the texture details of the HR directly from the LR.

The texture information is extracted by the GLCM method. GLCM represents the joint distribution of grayscale of two pixels with some spatial position relationship. The GLCM generation process as follows:

* A spot (x, y) and a spot $(x + a, y + b)$ in an image form a point pair. Let the pair of spots have a gray value of (f_1, f_2) and let the image have a gray value of at most L, then there will be a $L \times L$ combination of f_1 and f_2.
* For the whole image, the number of occurrences of each (f_1, f_2) value is calculated, and then they are arranged into a matrix.
* The total number of times (f_1, f_2) appears are normalized to obtain the probability $P_{(f_1, f_2)}$, which results in a grayscale co-generation matrix.

After obtaining the GLCM of the HR image, we utilize the properties of GLCM to extract the repeated texture details present in the image. Firstly, a 3×3 convolutional kernel is applied to expand the channels, followed by five multiscale residual blocks to explore features at different scales. In this part of the network, the texture feature map is upsampled to match the size of the HR image. Throughout this process, the LR image is used as the input to the network, allowing us to directly obtain the texture feature map of the LR image. The specific equations are as follows:

$$L_{texture} = \|T(I_{LR}) - I_{texture}\|_1 \tag{9}$$

where $T()$ denotes the texture reconstruction network, $I_{texture}$ represents the texture of the HR extracted by GLCM, $T(I_{LR})$ displays the reconstructed texture and make L_1 loss with $I_{texture}$

3.4   Fine Edge Reconstruction Network (FERN)

To address the issue of excessive false edges in the texture feature map generated by GLCM, we introduce the following edge extraction operator, which can produce more refined edge features:

$$I_{Edge} = div(u_h, u_v) \tag{10}$$

where $u_i = \frac{\nabla_i I_{HR}}{\sqrt{1 + |\nabla I_{HR}|^2}}$, $i \in \{h, v\}$, h and v represent two dimensions in different directions (horizontal and vertical), ∇ indicate gradient operation and $div()$ indicate the divergence operation. FERN has the same network structure as FTRN, but the target images and loss function utilized are different. The loss function of FERN is shown as:

$$L_{edge} = \|E(I_{LR}) - I_{edge}\|_1 \tag{11}$$

where the method of $E()$ stand for the fine edge reconstruction network, I_{edge} indicates the fine edge extracted by above methods, $E(I_{LR})$ displays the fine edge reconstructed by FERN and make L_1 loss with the fine edge detected by HR images.

3.5   Image Refinement Network (IRN)

For the image refinement module, we integrate the different features extracted from the aforementioned three branches. These three features enable us to capture sufficient texture details and obtain accurate edge features effectively. We fuse these three features and input them into the feature fine-tuning network. The entire process can be mathematically represented by the following equation:

$$f_{rb}^n = (w_2 * R(w_1 * f_{rb}^{n-1} + b_1) + b_2) + f_{rb}^{n-1} \qquad (12)$$

where w_i and b_i are commonly used weight parameters in neural networks of the layer i respectively, $i \in \{1, 2\}$, f_{rb}^n is the output of the nth residual block. $R()$ denotes the ReLU activation function.

In addition to the residual blocks, a long skip connection is used in these part to maintain the features at the input to the IRN and effectively limit the trouble of the disappearance of gradient. For the IRN last layer, we use a convolutional layer to convert the dimension to RGB channels, and then SR images are reconstructed. In training progress, L_1 loss function is used to reduce the gap between SR image and HR image.

$$L_{content} = \|I_{SR} - I_{HR}\|_1 \qquad (13)$$

In summary, we have designed a model called FETSR that can effectively reconstruct SR images. Typically, the edge and texture regions of an image contain abundant information that is challenging to reconstruct. Experimental results demonstrate that by incorporating fine-grained edge priors and texture priors, the reconstructed SR images exhibit accurate edges and rich texture details.

4   Experiments

4.1   Datasets

The DIV2K [1] is a common used dataset in super resolution reconstruction tasks, which contains 1000 images of various scenes, 800 of them are used for training, 100 can be used for validation, and 100 can be used for testing. As same as the previous works, we use a training dataset consisting of 800 images from DIV2K to train our model and meanwhile use the validation images from DIV2K to validate our model. During test our model, we employ the following datasets: Set5 [5], Set14 [34], BSDS100 [2] and Urban100 [13]. All of these test datasets are commonly used in super resolution and contain a variety of scenarios that are convincing enough to fully evaluate our model.

4.2   Implements Details

In training our network, we set the patch size to 48 as the input image block size and the batch size to 16, and constrain our training process by L_1 loss, edge loss, and texture loss, and set the weights to 1, 0.1, and 0.001, respectively. Since the pixel values of texture maps range from 0 to 255, we need to constrain it to the same dimension as the edge loss. The parameters of the optimizer are set to $\beta_1 = 0.9, \beta_2 = 0.999$, respectively, and the epoch is set to 600, and our residual block is finally set to 40. All code is based on the pytorch framework and is trained on 2 TITAN Xp GPUS.

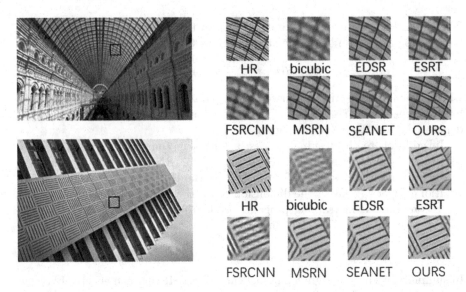

Fig. 3. A comparison of our model with other models shows that our model is able to reconstruct better visual effects and finer texture details.

4.3 Qualitative Comparisons and Discussion

As shown in the Fig. 3, we selected different images from the test dataset and reconstructed them with the available super-resolution. When compared with other super-resolution methods, our network is able to reconstruct not only more accurate texture information, but also sharper edges. In the first image, our reconstructed image has a better visual effect and a clearer reconstruction for some fine textures. In the second image, we reconstructed more accurate texture details. The third image clearly shows that the reconstructed image of our model highlights the edge line part of the floor.

4.4 Quantitative Comparisons and Discussion

As shown in Table 1, our model is compared with other neural network models. PSNR and SSIM are common metrics for judging the quality of reconstruction in super-resolution domains. Other methods have difficulty in reconstructing high quality images by learning the own features of LR images. Our model is able to effectively reconstruct the edges and textures of the images by using the prior generated from HR images, and is higher than other models in both PSNR and SSIM metrics.

5   Analysis and Discussion

5.1 Effectiveness of the Prior Information

It is a very important issue that how to use the effective prior information to aid the super-resolution reconstruction, so we conducted an experimental analysis of

Table 1. Quantitative analysis with existing CNN-based models for super-resolution reconstruction Highlighted indicates the best result.

Scale	Algorithm	Set5 PSNR↑/SSIM↑	Set14 PSNR↑/SSIM↑	BSDS100 PSNR↑/SSIM↑	Urban100 PSNR↑/SSIM↑
×2	SRCNN [33]	36.71/0.9536	32.32/0.9052	31.36/0.8880	29.54/0.8962
	FSRCNN [9]	37.06/0.9554	32.76/0.9078	31.53/0.8912	29.88/0.9024
	VDSR [15]	37.53/0.9583	33.05/0.9107	31.92/0.8965	30.79/0.9157
	SeaNet [11]	38.08/0.9609	33.75/0.9190	32.27/0.9008	32.50/0.9318
	Cross-SRN [20]	38.03/0.9606	33.62/0.9180	32.19/0.8997	32.28/0.9290
	MRFN [12]	37.98/0.9611	33.41/0.9159	32.14/0.8997	31.45/0.9221
	ESRT [21]	38.03/0.9600	33.75/0.9184	32.25/0.9001	32.58/0.9318
	FDSCSR [29]	38.12/0.9609	33.69/0.9191	32.24/0.9004	32.50/0.9315
	FETSR (ours)	**38.18/0.9612**	**33.90/0.9206**	**32.30/0.9010**	**32.68/0.9335**
×3	SRCNN [33]	32.47/0.9067	29.23/0.8201	28.31/0.7832	26.25/0.8028
	FSRCNN [9]	33.20/0.9149	29.54/0.8277	28.55/0.7945	26.48/0.8175
	VDSR [15]	33.68/0.9201	29.86/0.8312	28.83/0.7966	27.15/0.8315
	SeaNet [11]	34.55/0.9282	30.42/0.8444	29.17/0.8071	28.50/0.8594
	Cross-SRN [20]	34.43/0.9275	30.33/0.8417	29.09/0.8050	28.23/0.8535
	MRFN [12]	34.21/0.9267	30.03/0.8363	28.99/0.8029	27.53/0.8389
	ESRT [21]	34.42/0.9268	30.43/0.8433	29.15/0.8063	28.46/0.8574
	FDSCSR [29]	34.50/0.9281	30.43/0.8442	29.15/0.8068	28.40/0.8576
	FETSR (ours)	**34.61/0.9287**	**30.46/0.8454**	**29.22/0.8078**	**28.62/0.8615**
×4	SRCNN [33]	30.50/0.8573	27.62/0.7453	26.91/0.6994	24.53/0.7236
	FSRCNN [9]	30.73/0.8601	27.71/0.7488	26.98/0.7029	24.62/0.7272
	VDSR [15]	31.36/0.8796	28.11/0.7624	27.29/0.7167	25.18/0.7543
	SeaNet [11]	32.33/0.8970	28.72/0.7855	27.65/0.7388	26.32/0.7942
	Cross-SRN [20]	32.24/0.8954	28.59/0.7817	27.58/0.7364	26.16/0.7881
	MRFN [12]	31.90/0.8916	28.31/0.7746	27.43/0.7309	25.46/0.7654
	ESRT [21]	32.19/0.8947	28.69/0.7833	27.69/0.7379	26.39/0.7962
	FDSCSR [29]	32.36/0.8970	28.67/0.7840	27.63/0.7384	26.33/0.7935
	FETSR (ours)	**32.43/0.8978**	**28.74/0.7862**	**27.70/0.7405**	**26.41/0.7965**

Table 2. On scale x4, experimental results generated by different a prior information. Highlighted indicates the best result.

Edge	Texture	Set5 (PSNR↑/SSIM↑)	Set14 (PSNR↑/SSIM↑)	BSDS100 (PSNR↑/SSIM↑)
w/o	w/o	32.15/0.8952	28.55/0.7789	27.41/0.7368
w/o	w	32.32/0.8974	28.71/0.7861	27.67/0.7396
w	w/o	32.33/0.8970	28.72/0.7855	27.65/0.7388
w	w	**32.43/0.8978**	**28.74/0.7862**	**27.70/0.7405**

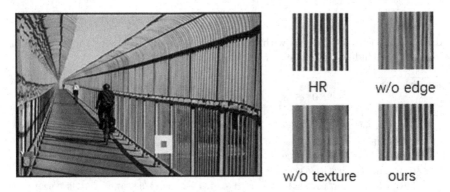

HR   w/o edge

w/o texture   ours

Fig. 4. Comparison chart of the ablation experiment, w/o indicates that the prior information is not introduced, w indicates that the prior information is introduced.

the factors affecting the super-resolution reconstruction. When we introduce only texture prior, we can see from Table 2 that the reconstruction of the image is not very good. If only fine edge prior information is introduced, the reconstruction effect is not very good for high frequency regions with regular pixel points and textures. As shown in Fig. 4, when we introduce both fine edge prior information and texture prior information, we can reconstruct the details of the image better.

Table 3. For the study of λ

λ	Set5 (PSNR↑/SSIM↑)
$\lambda_1 = 1, \lambda_2 = 1$	32.14/0.8874
$\lambda_1 = 0.1, \lambda_2 = 0.1$	32.31/0.8950
$\lambda_1 = 0.1, \lambda_2 = 0.01$	32.34/0.8962
$\lambda_1 = 0.1, \lambda_2 = 0.001$	**32.43/0.8978**

5.2   Study of λ

During the training process, the setting of hyper-parameters also has an important influence on the reconstruction effect. λ_1 and λ_2 are set to adjust the edge loss and the texture loss respectively. To weigh the influence of fine edges and textures in the reconstruction process, we set λ_1 to 0.1 and λ_2 to 0.001, thus controlling the texture loss and edge loss in the same dimension. From Table 3 We can see that different super parameter settings lead to different reconstruction effects.

6 Conclusion

In this article, we introduce a novel super-resolution network using fine edge and texture priors. The network consists of four components: a shallow feature extraction network, a fine texture reconstruction network, a fine edge reconstruction network, and an image refinement network. Our model uses fine edge and texture prior to not only reconstruct the internal texture details in the image, but also effectively avoid reconstructing the wrong edge information.

Acknowledgements. This work is supported by the National Natural Science Foundation of China (Nos. 62377029 and Nos. 22033002).

References

1. Agustsson, E., Timofte, R.: NTIRE 2017 challenge on single image super-resolution: dataset and study. In: Proceedings of the IEEE Conference on Computer Vision and Pattern Recognition Workshops, pp. 126–135 (2017)
2. Arbelaez, P., Maire, M., Fowlkes, C., Malik, J.: Contour detection and hierarchical image segmentation. IEEE Trans. Pattern Anal. Mach. Intell. **33**(5), 898–916 (2010)
3. Belekos, S.P., Galatsanos, N.P., Katsaggelos, A.K.: Maximum a posteriori video super-resolution using a new multichannel image prior. IEEE Trans. Image Process. **19**(6), 1451–1464 (2010)
4. Ben-Ezra, M., Zomet, A., Nayar, S.K.: Video super-resolution using controlled subpixel detector shifts. IEEE Trans. Pattern Anal. Mach. Intell. **27**(6), 977–987 (2005)
5. Bevilacqua, M., Roumy, A., Guillemot, C., Alberi-Morel, M.L.: Low-complexity single-image super-resolution based on nonnegative neighbor embedding (2012)
6. Canny, J.: A computational approach to edge detection. IEEE Trans. Pattern Anal. Mach. Intell. **6**, 679–698 (1986)
7. Chang, H., Yeung, D.Y., Xiong, Y.: Super-resolution through neighbor embedding. In: Proceedings of the 2004 IEEE Computer Society Conference on Computer Vision and Pattern Recognition, CVPR 2004, vol. 1, p. I. IEEE (2004)
8. Dai, T., Cai, J., Zhang, Y., Xia, S.T., Zhang, L.: Second-order attention network for single image super-resolution. In: Proceedings of the IEEE/CVF Conference on Computer Vision and Pattern Recognition, pp. 11065–11074 (2019)
9. Dong, C., Loy, C.C., Tang, X.: Accelerating the super-resolution convolutional neural network. In: Leibe, B., Matas, J., Sebe, N., Welling, M. (eds.) ECCV 2016. LNCS, vol. 9906, pp. 391–407. Springer, Cham (2016). https://doi.org/10.1007/978-3-319-46475-6_25
10. Duda, R.O., Hart, P.E., et al.: Pattern Classification and Scene Analysis, vol. 3. Wiley, New York (1973)
11. Fang, F., Li, J., Zeng, T.: Soft-edge assisted network for single image super-resolution. IEEE Trans. Image Process. **29**, 4656–4668 (2020)
12. He, Z., et al.: MRFN: multi-receptive-field network for fast and accurate single image super-resolution. IEEE Trans. Multimedia **22**(4), 1042–1054 (2019)
13. Huang, J.B., Singh, A., Ahuja, N.: Single image super-resolution from transformed self-exemplars. In: Proceedings of the IEEE Conference on Computer Vision and Pattern Recognition, pp. 5197–5206 (2015)

14. Hyun, C.M., Kim, H.P., Lee, S.M., Lee, S., Seo, J.K.: Deep learning for undersampled MRI reconstruction. Phys. Med. Biol. **63**(13), 135007 (2018)

15. Kim, J., Lee, J.K., Lee, K.M.: Accurate image super-resolution using very deep convolutional networks. In: Proceedings of the IEEE Conference on Computer Vision and Pattern Recognition, pp. 1646–1654 (2016)

16. Lee, D., Yoo, J., Tak, S., Ye, J.C.: Deep residual learning for accelerated MRI using magnitude and phase networks. IEEE Trans. Biomed. Eng. **65**(9), 1985–1995 (2018)

17. Li, F., Bai, H., Zhao, L., Zhao, Y.: Dual-streams edge driven encoder-decoder network for image super-resolution. IEEE Access **6**, 33421–33431 (2018)

18. Lim, B., Son, S., Kim, H., Nah, S., Mu Lee, K.: Enhanced deep residual networks for single image super-resolution. In: Proceedings of the IEEE Conference on Computer Vision and Pattern Recognition Workshops, pp. 136–144 (2017)

19. Liu, J., Zhang, W., Tang, Y., Tang, J., Wu, G.: Residual feature aggregation network for image super-resolution. In: Proceedings of the IEEE/CVF Conference on Computer Vision and Pattern Recognition, pp. 2359–2368 (2020)

20. Liu, Y., Jia, Q., Fan, X., Wang, S., Ma, S., Gao, W.: Cross-SRN: structure-preserving super-resolution network with cross convolution. IEEE Trans. Circ. Syst. Video Technol. **32**(8), 4927–4939 (2021)

21. Lu, Z., Li, J., Liu, H., Huang, C., Zhang, L., Zeng, T.: Transformer for single image super-resolution. In: Proceedings of the IEEE/CVF Conference on Computer Vision and Pattern Recognition, pp. 457–466 (2022)

22. Ren, S., Li, J., Tu, T., Peng, Y., Jiang, J.: Towards efficient video detection object super-resolution with deep fusion network for public safety. Secur. Commun. Netw. **2021**, 1–14 (2021)

23. Salvador, J., Perez-Pellitero, E.: Naive Bayes super-resolution forest. In: Proceedings of the IEEE International Conference on Computer Vision, pp. 325–333 (2015)

24. Sun, J., Xu, Z., Shum, H.Y.: Image super-resolution using gradient profile prior. In: 2008 IEEE Conference on Computer Vision and Pattern Recognition, pp. 1–8. IEEE (2008)

25. Tai, Y.W., Liu, S., Brown, M.S., Lin, S.: Super resolution using edge prior and single image detail synthesis. In: 2010 IEEE Computer Society Conference on Computer Vision and Pattern Recognition, pp. 2400–2407. IEEE (2010)

26. Tang, J., Zhang, J., Chen, D., Al-Nabhan, N., Huang, C.: Single-frame super-resolution for remote sensing images based on improved deep recursive residual network. EURASIP J. Image Video Process. **2021**(1), 1–19 (2021). https://doi.org/10.1186/s13640-021-00560-8

27. Timofte, R., De Smet, V., Van Gool, L.: Anchored neighborhood regression for fast example-based super-resolution. In: Proceedings of the IEEE International Conference on Computer Vision, pp. 1920–1927 (2013)

28. Wang, X., Yu, K., Dong, C., Loy, C.C.: Recovering realistic texture in image super-resolution by deep spatial feature transform. In: Proceedings of the IEEE Conference on Computer Vision and Pattern Recognition, pp. 606–615 (2018)

29. Wang, Z., Gao, G., Li, J., Yan, H., Zheng, H., Lu, H.: Lightweight feature de-redundancy and self-calibration network for efficient image super-resolution. ACM Trans. Multimed. Comput. Commun. Appl. **19**(3), 1–15 (2023)

30. Xie, J., Feris, R.S., Sun, M.T.: Edge-guided single depth image super resolution. IEEE Trans. Image Process. **25**(1), 428–438 (2015)

31. Yang, J., Wright, J., Huang, T.S., Ma, Y.: Image super-resolution via sparse representation. IEEE Trans. Image Process. **19**(11), 2861–2873 (2010)

32. Yang, W., et al.: Deep edge guided recurrent residual learning for image super-resolution. IEEE Trans. Image Process. **26**(12), 5895–5907 (2017)
33. Yoon, Y., Jeon, H.G., Yoo, D., Lee, J.Y., So Kweon, I.: Learning a deep convolutional network for light-field image super-resolution. In: Proceedings of the IEEE International Conference on Computer Vision Workshops, pp. 24–32 (2015)
34. Zeyde, R., Elad, M., Protter, M.: On single image scale-up using sparse-representations. In: Cohen, A., et al. (eds.) Curves and Surfaces 2010. LNCS, vol. 6920, pp. 711–730. Springer, Heidelberg (2012). https://doi.org/10.1007/978-3-642-27413-8_47
35. Zhang, Y., Tian, Y., Kong, Y., Zhong, B., Fu, Y.: Residual dense network for image super-resolution. In: Proceedings of the IEEE Conference on Computer Vision and Pattern Recognition, pp. 2472–2481 (2018)
36. Zhou, Q., Chen, S., Liu, J., Tang, X.: Edge-preserving single image super-resolution. In: Proceedings of the 19th ACM International Conference on Multimedia, pp. 1037–1040 (2011)

UD-GCN: Uncertainty-Based Semi-supervised Deep GCN for Imbalanced Node Classification

Baifan Wei[1,2] and Qing He[1,2,3(✉)]

[1] Henan Institute of Advanced Technology, Zhengzhou University, Zhengzhou 450052, People's Republic of China
weibaifan@gs.zzu.edu.cn
[2] Key Lab of Intelligent Information Processing of Chinese Academy of Sciences (CAS), Institute of Computing Technology, CAS, Beijing 100190, China
heqing@ict.ac.cn
[3] University of Chinese Academy of Sciences, Beijing 100049, China

Abstract. Fraud detection, especially graph-based approaches, has received increasing attention in recent years because graph structures can fully represent the information about the relationships between data, which facilitates fraudster detection. However, many real-world applications have encountered difficulties in classifying in unbalanced data, which is common in areas such as financial fraud. We argue that the key to this difficulty is not only the data itself, i.e., the unbalanced number of samples from different classes, but equally stems from the fading of minority class node features by majority class neighbors due to the graph neural network neighborhood aggregation mechanism. We propose a semi-supervised deep graph convolutional classification model with unbalanced graph nodes that aims to address this challenge by exploiting the impact of the under-sampled adaptive coordination of the classifier's own features on model performance, which is integrated in a deep graph convolutional network that contains multiple layers of the same simplified graph network architecture and a nonlinear function that can be recursively optimized. Extensive experiments show that our approach still yields robust performance even in highly imbalanced graphs with sparse labels.

Keywords: Class imbalance · Data re-sampling · Graph neural network · Fraud detection

1 Introduction

1.1 Introduction

Fraud detection is a widely used task with far-reaching implications in various fields such as security [27], finance [6,19,36,41], healthcare [9,12,15,30], and censorship [7,22,29]. While a number of techniques have been proposed to

E. Chen et al. (Eds.): BigData 2023, CCIS 2005, pp. 112–124, 2023.
https://doi.org/10.1007/978-981-99-8979-9_9

detect fraudsters in multidimensional point sets, fraud detection [1,3,28] based on graph-structured data has recently received increasing attention as graph data becomes ubiquitous [42]. In essence, the underlying assumption of graph-based fraud detection is homogeneous preference, i.e., similarity between nodes and their neighbors. This is in line with our real-world intuition that similar behaviors will exist between fraudsters and other fraudsters, which in turn leads to certain connections that can be made to identify other fraudsters through one fraudster.

However, in the vast majority of fraud detection tasks, the number of nodes in different categories can vary significantly. For instance, in the real-world review dataset from YelpChi [29], the ratio of spam reviews to benign reviews is 6:1. When a fraudulent user's neighbors are mostly good people and only individually fraudulent, GNN, whose core idea is message passing, will be disturbed by neighbors passing messages and then incorrectly identify this fraudster as a good person because his neighbors are mostly good people [39]. Furthermore, real-world fraud detection tasks are highly likely to encounter other difficulties, such as fraudsters masquerading as good guys [7], which can cause noise to be trapped in the dataset, and this noise can interfere with the detection of fraudsters. Spammers, for example, typically use benign accounts for their spam comments to occur in order to hide themselves among benign users. This highly unbalanced, noisy and falsely labeled data poses a serious challenge to the downstream classification task.

Traditional research on category imbalance problems has been directed in two main directions: data-level approaches that preprocess the dataset to balance the distribution, and algorithm-level approaches that modify the loss function for cost-sensitive adjustment. Most of the imbalanced classification problems on graph structure have also followed these two directions:

Data-level methods modify the dataset to balance the sample distribution based on the number of samples in different categories to make it suitable for standard learning algorithms, and consist of three methods: under-sampling methods that reduce the number of samples in the majority category [34], over-sampling methods that generate new minority samples to increase the number of samples in the minority category [40], and hybrid methods that simultaneously correcting the distribution of minority and majority class samples [21]. However, the data-level methods on graphs are based on distance-based and labeling designs, and they are not suitable for application to datasets with missing values or false labels, which are common in realistic datasets. This is because fraudsters may be at a very close Euclidean distance from their benign neighbors. In addition, oversampling methods may suffer from large computational costs when applied to large-scale data.

Algorithm-level methods modify existing classification algorithms, such as loss functions, in accordance with a priori knowledge to assign corresponding weights to different samples to mitigate the classification boundaries that favor majority classes. However, by cost-sensitive tuning methods [4,5,18] when working with batch-trained classifiers, they may fail on highly imbalanced datasets. Because they cannot mitigate the problem of too small a proportion of minority class samples, they may result in minority class samples being included in only a few batches, which can quickly drive the model into local minima. Note also that task-specific cost matrices are usually given in advance by domain experts, which is not available in many real-world problems.

In summary, none of the mainstream methods currently available can handle well the classification task on highly imbalanced graphs with missing values or false labels, which is a common problem encountered in real-world fraud detection tasks. In addition to the reasons mentioned above, another major reason for the failure of existing methods in is that they ignore the difficulty of imbalanced classification tasks, not only stemming from the characteristics of the samples themselves, such as the presence of noisy samples [26], the overlapping underlying distribution between classes [10], and the distribution gap of sample size in different categories, but also the influence of the characteristics of the classifier itself, such as the aggregation of messages in GNNs [13,14,35] may lead to the minority class adjacent to the majority class features are faded by the majority class neighbors as a result.

Recent research states that to determine whether a node is homogeneous or heterogeneous, the concept of node-level homogeneity ratio is defined as the fraction of neighbors with the same class as the node as a proportion of all neighbors of the node. Most GNNs work well on strongly homogeneous nodes, but fail on strongly heterogeneous nodes [20]. The performance gap between graph neural networks working on homogeneous and heterogeneous nodes will result in fraud detection tasks suffering from significant challenges. As mentioned before, this is because the core mechanism of GNNs is neighborhood aggregation, but in an imbalanced setting, most of the neighbors of the fraudster as a minority class are likely to be benign nodes as a majority class. As a result, the features of the fraudsters themselves are easily ignored by the classifier and the prediction results are dominated by the benign neighbors. We can plan the sampling strategy accordingly to mitigate the misleading effect of heterogeneous neighborhoods of strongly heterogeneous nodes on the classifier. The degree of heterogeneity of nodes is measured by the uncertainty of classifier predictions rather than the class labels of nodes, since the predictions of heterogeneous nodes will show greater variance for sparsely labeled datasets. Furthermore, the uncertainty of the prediction results naturally fits the model used for classification as it is defined relative to the given classifier.

Based on the uncertainty in the prediction of heterogeneous nodes, this paper proposes a new learning framework named UD-GCN. Instead of simply and directly balancing the number of samples from different classes or assigning weights based on labels or distances, we consider the performance differences of the classifier on different samples and iteratively sample those samples that can contribute more to the model and provide more favorable information based on the uncertainty in the prediction results. Majority class samples that provide more favorable information. The under-sampling strategy is integrated into a deep graph convolutional neural network that contains many layers of a structurally identical simplified graph convolutional network and a nonlinear function capable of recursive optimization. This adaptive iterative process allows our framework to gradually focus on the more difficult to classify strongly heterogeneous samples, while still retaining the knowledge learned from the strongly homogeneous samples to prevent over-fitting.

In summary, the contributions of this paper are as follows.

- This paper demonstrates the reasons why traditional imbalance learning methods fail in real-world imbalanced graph classification tasks, which are informative for other similar classification problems.
- We propose UD-GCN, a semi-supervised learning framework for unbalanced data classification. By considering the performance gap of classifiers on samples with different degrees of heterogeneity, our proposed model is automatically optimized in a classifier-specific manner. Compared with existing methods, UD-GCN is accurate, robust, and adaptive.
- Unlike the mainstream approaches, our model does not require any predefined distance metric or other prior knowledge, which is usually difficult to obtain in real-world problems.

2 Methodology

In this section, we introduce the proposed UD-GCN model framework. Sections 2.1 and 2.2 detail the adaptive sample sampling strategy based on model uncertainty and the recursive optimization of the deep graph convolution model, respectively, with iterative details as illustrated in Fig. 1. And the whole algorithm is summarized in Sect. 2.3.

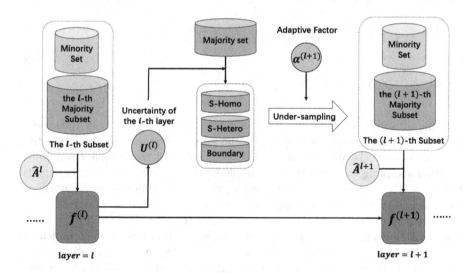

Fig. 1. In each iteration, most classes in the dataset with small uncertainty scores will be partially discarded, and their discard probability is determined by the adaptive factor α, a value that increases as the number of layers increases. The sampled new subset is fed into the l-th layer for learning along with $\hat{A}$ to the $l-1$-th power.

2.1   Adaptive Under-Sampling

Since the features of those nodes connected to non-identical nodes themselves are easily diluted by the information passed from their neighbors, the classification results of GNNs may exhibit higher uncertainty relative to homogeneous nodes for heterogeneous nodes. We can determine whether a node is biased homogeneous or heterogeneous by capturing the uncertainty of the model. Specifically, given an undirected graph $\mathcal{G} = (X, \hat{A})$ with N nodes and a prediction $\hat{Y}$ such that the parameter W_b is a random variable obeying some prior distribution (e.g., Gaussian distribution $W \sim \mathcal{N}(0, I)$), the likelihood function of the node classification model is Eq. (1).

$$p(\hat{Y}|\mathcal{V}, \mathcal{E}) = \int_W p(\hat{Y}|W, X, \hat{A})p(W|X, \hat{A})dW \tag{1}$$

To obtain the posterior probability distribution $p(W|X, \hat{A}$ in Eq. (1), we choose to approximate the solution using Monte Carlo dropout variational inference [8]. The loss function is defined as Eq. (2).

$$L(W_b) = -\frac{1}{T}\sum_{t=1}^{T} Y\log(\hat{Y}_t) + \frac{1-\theta}{2T}\|W_b\|^2 \tag{2}$$

where $\hat{Y}_t = f_{\hat{W}_t}(X, \hat{A})$ is the output after multiple sampling of different W_b and θ is the hyper-parameter. After training, we obtain from Eq. (3) an N-dimensional vector U used to measure the uncertainty level of the model, which represents the uncertainty score of each node.

$$\mathbb{E}[\hat{Y}|X, \hat{A}] = \int \hat{Y}p(\hat{Y}|X, \hat{A})d\hat{Y} \approx \frac{1}{T}\sum_{t=1}^{T} \hat{Y}_t$$
$$U[\hat{Y}|X, \hat{A}] = \text{Var}(\hat{Y}|X, \hat{A}) \approx \frac{1}{T}\sum_{t=1}^{T}(\hat{Y}_t - \mathbb{E}[\hat{Y}|X, \hat{A}])^2 \tag{3}$$

   The advantage of using sample-predicted uncertainty scores in the case of unbalanced classification is that it fills the gap between the sampling strategy used to balance the number of sample categories and the classifier capability. Existing sampling methods almost completely ignore the classifier's capability. However, GNN-based classifiers typically exhibit very different performance on data samples with different degrees of heterogeneity, even if they have the same imbalance rate, so the imbalance rate does not reflect well the difficulty of the classification task. In contrast, the uncertainty scores give a good indication of the task difficulty of the chosen classifier on different samples, allowing our framework to optimize the classification performance of the model in a classifier-specific manner. In particular, using the uncertainty level of sample predictions, rather than the imbalance rate, reduces the reliance on sample labels, allowing our framework to be applicable to scenarios with sparse samples and spurious labels.

Intuitively, data samples can be classified into three categories based on their corresponding uncertainty scores, i.e., strongly homogeneous samples, strongly heterogeneous samples and boundary samples: homogeneous samples can be well classified by GNN models, so each sample contributes a small uncertainty score. However, due to their large number, their overall contribution cannot be ignored. For such samples, we only need to keep a small portion as a skeleton to represent the whole to prevent over-fitting, and then discard most of them because they have been well learned by the classifier. In contrast, there are also strongly heterogeneous samples, each with a large degree of prediction uncertainty even though their number is usually small. As a result, the sum of uncertainty scores can be very large. Such samples are usually caused by prominent outliers or outliers, and forcing the classifier to learn such samples may lead to severe over-fitting. For the other samples, they can be regarded as boundary samples, which are in the middle range of heterogeneity and therefore mostly located on the decision boundary. The boundary samples are the samples that provide the most effective information during training. Having the classifier learn more about these boundary samples usually helps to further improve the final classification performance.

Inspired by the previous paper, our goal is to design an under-sampling strategy that balances the number of samples from different classes while reducing the influence of strongly homogeneous and strongly heterogeneous samples on the classifier and increasing the importance of boundary samples containing higher amounts of valid information.

In the initial stage of learning, the three sets of samples are downsampled with the same probability, which is to gently reduce the number of majority samples and reduce the negative impact of outlier samples on the performance of the classifier, in order to allow the classifier to focus more on the boundary samples [16]. As the classifier gradually adapts to the dataset during the learning process, too many strongly homogeneous samples will prevent the model from learning in later iterations because they provide less information. To this end, an adaptive factor α is introduced to coordinate the sampling probability, whose value gradually increases with iteration, increasing the sampling probability for samples with higher uncertainty scores and decreasing the sampling probability for samples with lower uncertainty scores and smaller contributions, allowing the classifier to learn those strongly heterogeneous samples that are difficult to classify in order to improve the classification performance. In Fig. 2, the details of the work of the adaptive factor α are shown with the Cora [31] data set.

In summary, in the early stage of the iteration, our model mainly learns the boundary samples. In the later part of the iteration, our framework focuses on the strongly heterogeneous samples while retaining the information learned from the boundary samples in the earlier stages, which effectively prevents the occurrence of over-fitting.

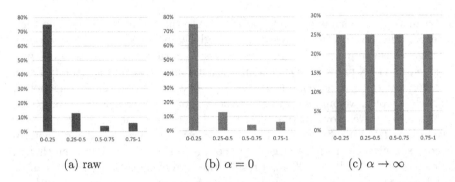

(a) raw   (b) $\alpha = 0$ (c) $\alpha \rightarrow \infty$

Fig. 2. The Cora data set is used as an example to demonstrate how the adaptive factor α controls under-sampling. Each subplot represents the percentage of different uncertainty scores in the dataset. Subplot (a) is the original distribution of all majority class samples. Subplot (b) is the distribution of the subset of majority class samples obtained by sampling when $\alpha = 0$. Subplot (c) is the distribution of the subset of majority class samples sampled when $\alpha \rightarrow \infty$.

2.2   Recursive Optimization for Deep GCN

The graph embedding of the nodes of a semi-supervised node classification vanilla GCN model with two convolutional layers [14] is given by Eq. (4).

$$Z = \hat{A}\text{ReLU}(\hat{A}XW^{(0)})W^{(1)} \tag{4}$$

where $Z \in \mathbb{R}^{N \times K}$ is the node embedding matrix, N is the number of nodes, and K is the number of categories. $X \in \mathbb{R}^{N \times C}$ is the feature matrix, where C is the feature dimension. $\hat{A} = \tilde{D}^{-\frac{1}{2}}\tilde{A}\tilde{D}^{-\frac{1}{2}}$, where $\tilde{A} = A + I$, $A \in \mathbb{R}^{N \times N}$ is the adjacency matrix of the undirected graph $\mathcal{G}$, $\tilde{D}$ is the degree matrix of $\tilde{A}$. In addition, $W^{(0)} \in \mathbb{R}^{C \times H}$ and $W^{(1)} \in \mathbb{R}^{H \times K}$ are the weight matrices of the two layers respectively.

The purpose of building deep graph neural networks is to cooperate with sampling algorithms to adaptively fill the gap between classifier performance and data characteristics, while efficiently exploring information from higher-order neighbors. However, if the number of layers of the graph convolutional network is increased directly, the local information in the nodes will be lost because they are smoothed [23]. Fortunately, obtaining an adequate representation of node features does not require much nonlinear transformation because the features of each node are usually one-dimensional, rather than multidimensional data structures, and two-layer fully connected neural network is a more convenient option [11,37,38]. Similarly, we follow this direction. First, to reduce the optimization cost of the multi-parameter matrix, remove the ReLU [37] and simplify the embedding to Eq. (5):

$$Z = \hat{A}^l XW^{(0)} \dots W^{(l-1)} = \hat{A}^l X\tilde{W} \tag{5}$$

where $W^{(0)} \dots W^{(l-1)}$ is collapsed to $\tilde{W}$ and $\hat{A}^l$ denotes the lth power of $\hat{A}$. In particular, removing the ReLU operation also alleviates the over-smoothing

---

Algorithm 1: UD-GCN

Input: Training set $\mathcal{G} = (X, \hat{A})$, a two-layer simplified graph convolution f and the number of layers L

Output: Final prediction $F(X, \hat{A})$

initialization: $\mathcal{P} \Leftarrow$ minority in $\mathcal{G}$, $\mathcal{N} \Leftarrow$ majority in $\mathcal{G}$, $\hat{X}^{(0)} = X$ and random under-sample majority subsets $\mathcal{N}^{(0)}$ from $\mathcal{N}$, where $|\mathcal{N}^{(0)}| = |\mathcal{P}|$

for l=0 **to** L **do**

> Train the $f^{(l)}$ using $\hat{X}^{(l)}$, $\mathcal{P}$ and $\mathcal{N}^{(l)}$
>
> Evaluate the uncertainty of the current layer model $U^{(l)}$ using. Eq.(3)
>
> Update the feature matrix used in the next layer $\hat{X}^{(l+1)} = \hat{A}\hat{X}^{(l)}$
>
> Update adaptive factor $\alpha = \tan(\frac{l\pi}{2L})$ and sampling weight $p^{(l)} = \frac{1}{U^{(l)}+\alpha}$
>
> Under-sample majority subsets $\mathcal{N}^{(l+1)}$ from $\mathcal{N}$ using $p^{(l)}$, where $|\mathcal{N}^{(l+1)}| = |\mathcal{P}|$

end

return $F(X, \hat{A}) = f^{(L)}$

---

problem of node embedding [17]. However, this linear stacking transformation of the graph convolution has limited ability to learn higher-order neighbor information. Using a suitable nonlinear function f to replace $\tilde{W}$ of Eq. (5) [2,33], the final embedding matrix of the l-th layer is obtained as Eq. 6, and the l-th hop neighbor feature of the current node can be learned.

$$Z^{(l)} = f(\hat{A}^l X) \tag{6}$$

where $\hat{A}^l X = \hat{A} \cdot (\hat{A}^{l-1})X$. In particular, f is nonlinear rather than linear. It will be recursively optimized in each layer of the model so that each layer is initialized using the parameters of the previous layer to guarantee the representation power.

2.3 Algorithm Formalization

Finally, our algorithm is formally described in this subsection, and the details are given in Algorithm 1. Note that the uncertainty score is updated in each iteration and the tan function is used to adjust the change in the adaptive factor α in order to select the sample that contains the most valid information for the current stage.

3 Experiments

3.1 Experimental Setup

Datasets. The commonly used graph datasets Plantoid paper citation graph (Cora, CiteSeer, Pubmed) [31] and Amazon co-purchase graph (Photo, Computers) [25] were chosen to validate the effectiveness of our method. The dataset is also modified by adjusting the imbalance rate IR to 10 and the tagging rate ρ to 5% to construct application scenarios with high imbalance and scarce tags.

Compared Methods. Our approach is compared with three classical GNN algorithms (GCN [14], GAT [35], GraphSAGE [13], and two imbalanced algorithms applied on graphs (DR-GCN [32] and GraphSMOTE [40] to validate the performance of our framework in fraud detection tasks on imbalanced graphs.

Metrics. Metrics used to evaluate model performance on classification tasks on unbalanced datasets should not be biased towards any class of nodes [24]. Since accuracy does not reflect model performance well, F1-macro (the unweighted average of F1 scores for each category) was used for evaluation, with higher scores for the metrics indicating higher performance of these methods.

3.2   Performance Comparison

To test the classification performance of our model on a highly unbalanced and sparsely labeled dataset, it is compared with all comparative methods. The results in Table 1 show that our method outperforms all the baselines. To reduce the effect of randomness, the mean and standard deviation of five independent runs were performed.

Table 1. Performance comparison for node classification on Imbalance graphs. The best results are bolded.

Dataset	Cora	CiteSeer	PubMed	Photo	Computer
GCN	49.2 ± 1.5	44.8 ± 1.2	51.7 ± 1.8	43.4 ± 1.6	39.8 ± 2.6
GAT	46.9 ± 1.6	44.9 ± 2.0	49.2 ± 2.1	47.3 ± 2.4	41.4 ± 2.3
GraphSAGE	$43.5 + 1.2$	51.2 ± 1.5	53.4 ± 1.7	43.8 ± 2.2	39.2 ± 1.9
DR-GCN	51.2 ± 1.3	52.5 ± 2.6	77.2 ± 2.3	78.5 ± 2.9	64.4 ± 3.1
GraphSMOTE	49.6 ± 1.1	51.8 ± 1.3	75.4 ± 1.9	77.5 ± 2.3	59.3 ± 2.4
UD-GCN	$\mathbf{53.2 \pm 1.6}$	$\mathbf{54.7 \pm 2.1}$	$\mathbf{78.2 \pm 2.1}$	$\mathbf{80.2 \pm 2.2}$	$\mathbf{65.4 \pm 2.8}$

The results show that even in the case of a highly unbalanced dataset (IR = 10) with sparse labels ($\rho = 5\%$), our method performs well, with evaluation scores significantly better than the other baselines. Specifically, the methods used for comparison are discussed in two groups. The first group is GCN, GAT and GraphSAGE, and the second group is DR-GCN and GraphSMOTE.

The performance of the first group of methods is the worst of all the methods used for comparison because they do not have a mechanism for solving the imbalance problem, resulting in a few classes not being learned sufficiently. Among them, GAT, which performs better on balanced graphs, instead becomes the most failed presence among these three methods on highly unbalanced datasets, because the attention mechanism can only fully perform if it contains more available information, but highly unbalanced datasets lack enough minority class data for the model to learn to obtain more accurate output results.

The second group, DR-GCN and GraphSMOTE are methods designed specifically for unbalanced graphs and are one of the representatives of algorithm-level methods and data-level methods, respectively. This group of methods has improved performance relative to the generic graph neural network methods, however, DR-GCN also does not balance the number of different labeled samples, which may cause the nonconvex optimization process using gradient descent update to quickly fall into local extrema. Although GraphSMOTE balances the class distribution, its distance- and label-based oversampling strategy is not adapted to label-sparse datasets, and its sampling strategy is completely independent of the chosen classifier and cannot obtain samples that are more useful for the classifier. Therefore both have limited improvement on model performance.

3.3   Sensitivity to the Number of Model Layers

UD-GCN has one key hyper-parameter: the number of basic classifier layers l. As discussed earlier, the stacking of graph convolutions tends to degrade performance by subjecting node embedding to over-smoothing [17]. However, our deep graph convolution model removes the redundant design to avoid the over-smoothing problem. This subsection conducts experiments on the dataset to verify the effect of the number of classifier layers l on the predictive performance of our model and to test whether this anti-transition smoothing mechanism is effective. The results are shown in Fig. 3. It can be observed that UD-GCN is able to improve the performance on all three datasets as the number of layers increases without performance degradation due to transition smoothing.

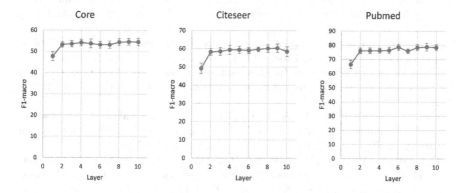

Fig. 3. The F1-macro scores of UD-GCN for different number of layers with error bars as standard deviations for five independent runs. It is quite straightforward to observe that UD-GCN is able to improve the performance of all three datasets as the number of layers increases without any degradation in performance due to transition smoothing.

4   Conclusion

In this paper, we propose a new semi-supervised node classification learning model for highly unbalanced and sparsely labeled graphs, UD-GCN.We argue that in addition to the data itself, such as noise, the number of samples in different categories, the performance gap exhibited by GNN's ability to aggregate neighborhood information on nodes with different degrees of heterogeneity is crucial for performance of the fraud detection task is equally critical. Therefore, we introduce model prediction uncertainty as a metric for down-sampling the dataset to fill the gap between sampling strategy and classifier performance. Our model is tested in the context of a highly unbalanced and label-sparse task. It yields higher performance compared to other methods. Overall, incorporating the characteristics of the model itself into the consideration of the data pre-processing approach may be a promising direction for similar research.

Acknowledgements. The research work supported by National Key R&D Plan No. 2022YFC3303302, the National Natural Science Foundation of China under Grant No.61976204. This work was also supported by the National Natural Science Foundation of China (No. 31900979, U1811461) and Zhengzhou Collaborative Innovation Major Project (No. 20XTZX11020).

References

1. Akoglu, L., Tong, H., Koutra, D.: Graph based anomaly detection and description: a survey. Data Min. Knowl. Disc. **29**, 626–688 (2015). https://doi.org/10.1007/s10618-014-0365-y
2. Belilovsky, E., Eickenberg, M., Oyallon, E.: Greedy layerwise learning can scale to ImageNet. In: International Conference on Machine Learning, pp. 583–593. PMLR (2019)
3. Boniol, P., Palpanas, T., Meftah, M., Remy, E.: GraphAn: graph-based subsequence anomaly detection. Proc. VLDB Endow. **13**(12), 2941–2944 (2020)
4. Chen, D., et al.: Topology-imbalance learning for semi-supervised node classification. In: Advances in Neural Information Processing Systems, vol. 34, pp. 29885–29897 (2021)
5. Cui, Y., Jia, M., Lin, T.Y., Song, Y., Belongie, S.: Class-balanced loss based on effective number of samples. In: Proceedings of the IEEE/CVF Conference on Computer Vision and Pattern Recognition, pp. 9268–9277 (2019)
6. Dal Pozzolo, A., Boracchi, G., Caelen, O., Alippi, C., Bontempi, G.: Credit card fraud detection: a realistic modeling and a novel learning strategy. IEEE Trans. Neural Netw. Learn. Syst. **29**(8), 3784–3797 (2017)
7. Dou, Y., Liu, Z., Sun, L., Deng, Y., Peng, H., Yu, P.S.: Enhancing graph neural network-based fraud detectors against camouflaged fraudsters. In: Proceedings of the 29th ACM International Conference on Information & Knowledge Management, pp. 315–324 (2020)
8. Gal, Y., Ghahramani, Z.: Dropout as a Bayesian approximation: representing model uncertainty in deep learning. In: International Conference on Machine Learning, pp. 1050–1059. PMLR (2016)

9. Gamberger, D., Lavrac, N., Groselj, C.: Experiments with noise filtering in a medical domain. In: ICML, vol. 99, pp. 143–151 (1999)

10. García, V., Sánchez, J., Mollineda, R.: An empirical study of the behavior of classifiers on imbalanced and overlapped data sets. In: Rueda, L., Mery, D., Kittler, J. (eds.) CIARP 2007. LNCS, vol. 4756, pp. 397–406. Springer, Heidelberg (2007). https://doi.org/10.1007/978-3-540-76725-1_42

11. Gasteiger, J., Bojchevski, A., Günnemann, S.: Predict then propagate: graph neural networks meet personalized pagerank. arXiv preprint arXiv:1810.05997 (2018)

12. Haixiang, G., Yijing, L., Shang, J., Mingyun, G., Yuanyue, H., Bing, G.: Learning from class-imbalanced data: review of methods and applications. Expert Syst. Appl. **73**, 220–239 (2017)

13. Hamilton, W., Ying, Z., Leskovec, J.: Inductive representation learning on large graphs. In: Advances in Neural Information Processing Systems, vol. 30 (2017)

14. Kipf, T.N., Welling, M.: Semi-supervised classification with graph convolutional networks. arXiv preprint arXiv:1609.02907 (2016)

15. Kumar, M., Ghani, R., Mei, Z.S.: Data mining to predict and prevent errors in health insurance claims processing. In: Proceedings of the 16th ACM SIGKDD International Conference on Knowledge Discovery and Data Mining, pp. 65–74 (2010)

16. Li, B., Liu, Y., Wang, X.: Gradient harmonized single-stage detector. In: Proceedings of the AAAI Conference on Artificial Intelligence, vol. 33, pp. 8577–8584 (2019)

17. Li, Q., Han, Z., Wu, X.M.: Deeper insights into graph convolutional networks for semi-supervised learning. In: Proceedings of the AAAI Conference on Artificial Intelligence, vol. 32 (2018)

18. Lin, T.Y., Goyal, P., Girshick, R., He, K., Dollár, P.: Focal loss for dense object detection. In: Proceedings of the IEEE International Conference on Computer Vision, pp. 2980–2988 (2017)

19. Liu, C., et al.: Fraud transactions detection via behavior tree with local intention calibration. In: Proceedings of the 26th ACM SIGKDD International Conference on Knowledge Discovery & Data Mining, pp. 3035–3043 (2020)

20. Liu, Y., Ao, X., Feng, F., He, Q.: UD-GNN: uncertainty-aware debiased training on semi-homophilous graphs. In: Proceedings of the 28th ACM SIGKDD Conference on Knowledge Discovery and Data Mining, pp. 1131–1140 (2022)

21. Liu, Y., et al.: Pick and choose: a GNN-based imbalanced learning approach for fraud detection. In: Proceedings of the Web Conference 2021, pp. 3168–3177 (2021)

22. Liu, Z., Dou, Y., Yu, P.S., Deng, Y., Peng, H.: Alleviating the inconsistency problem of applying graph neural network to fraud detection. In: Proceedings of the 43rd International ACM SIGIR Conference on Research and Development in Information Retrieval, pp. 1569–1572 (2020)

23. Luan, S., Zhao, M., Chang, X.W., Precup, D.: Break the ceiling: stronger multiscale deep graph convolutional networks. In: Advances in Neural Information Processing Systems, vol. 32 (2019)

24. Luque, A., Carrasco, A., Martín, A., de Las Heras, A.: The impact of class imbalance in classification performance metrics based on the binary confusion matrix. Pattern Recogn. **91**, 216–231 (2019)

25. McAuley, J., Targett, C., Shi, Q., Van Den Hengel, A.: Image-based recommendations on styles and substitutes. In: Proceedings of the 38th International ACM SIGIR Conference on Research and Development in Information Retrieval, pp. 43–52 (2015)

26. Napierała, K., Stefanowski, J., Wilk, S.: Learning from imbalanced data in presence of noisy and borderline examples. In: Szczuka, M., Kryszkiewicz, M., Ramanna, S., Jensen, R., Hu, Q. (eds.) RSCTC 2010. LNCS (LNAI), vol. 6086, pp. 158–167. Springer, Heidelberg (2010). https://doi.org/10.1007/978-3-642-13529-3_18

27. Neville, J., Şimşek, Ö., Jensen, D., Komoroske, J., Palmer, K., Goldberg, H.: Using relational knowledge discovery to prevent securities fraud. In: Proceedings of the Eleventh ACM SIGKDD International Conference on Knowledge Discovery in Data Mining, pp. 449–458 (2005)

28. Pourhabibi, T., Ong, K.L., Kam, B.H., Boo, Y.L.: Fraud detection: a systematic literature review of graph-based anomaly detection approaches. Decis. Support Syst. **133**, 113303 (2020)

29. Rayana, S., Akoglu, L.: Collective opinion spam detection: bridging review networks and metadata. In: Proceedings of the 21th ACM SIGKDD International Conference on Knowledge Discovery and Data Mining, pp. 985–994 (2015)

30. Sariyar, M., Borg, A., Pommerening, K.: Controlling false match rates in record linkage using extreme value theory. J. Biomed. Inform. **44**(4), 648–654 (2011)

31. Sen, P., Namata, G., Bilgic, M., Getoor, L., Galligher, B., Eliassi-Rad, T.: Collective classification in network data. AI Mag. **29**(3), 93–93 (2008)

32. Shi, M., Tang, Y., Zhu, X., Wilson, D., Liu, J.: Multi-class imbalanced graph convolutional network learning. In: Proceedings of the Twenty-Ninth International Joint Conference on Artificial Intelligence (IJCAI-2020) (2020)

33. Sun, K., Zhu, Z., Lin, Z.: AdaGCN: adaboosting graph convolutional networks into deep models. arXiv preprint arXiv:1908.05081 (2019)

34. Tomek, I.: Two modifications of CNN. IEEE Trans. Syst. Man Cybern. **SMC-6**(11), 769–772 (1976)

35. Veličković, P., Cucurull, G., Casanova, A., Romero, A., Lio, P., Bengio, Y.: Graph attention networks. arXiv preprint arXiv:1710.10903 (2017)

36. Wang, D., et al.: A semi-supervised graph attentive network for financial fraud detection. In: 2019 IEEE International Conference on Data Mining (ICDM), pp. 598–607. IEEE (2019)

37. Wu, F., Souza, A., Zhang, T., Fifty, C., Yu, T., Weinberger, K.: Simplifying graph convolutional networks. In: International Conference on Machine Learning, pp. 6861–6871. PMLR (2019)

38. Xu, K., Hu, W., Leskovec, J., Jegelka, S.: How powerful are graph neural networks? arXiv preprint arXiv:1810.00826 (2018)

39. Yang, Y., Xu, Z.: Rethinking the value of labels for improving class-imbalanced learning. In: Advances in Neural Information Processing Systems, vol. 33, pp. 19290–19301 (2020)

40. Zhao, T., Zhang, X., Wang, S.: GraphSMOTE: imbalanced node classification on graphs with graph neural networks. In: Proceedings of the 14th ACM International Conference on Web Search and Data Mining, pp. 833–841 (2021)

41. Zhong, Q., et al.: Financial defaulter detection on online credit payment via multi-view attributed heterogeneous information network. In: Proceedings of the Web Conference 2020, pp. 785–795 (2020)

42. Zhou, J., et al.: Graph neural networks: a review of methods and applications. AI Open **1**, 57–81 (2020)

Twin Support Vector Regression with Privileged Information

Yanmeng Li[1] and Wenzhu Yan[2($\boxtimes$)]

[1] School of Internet of Things, Nanjing University of Posts and Telecommunications,
Nanjing 210003, China
liyanmeng@njupt.edu.cn
[2] School of Computer and Electronic Information, Nanjing Normal University,
Nanjing 210023, China
yanwenzhu@nnu.edu.cn

Abstract. In this paper, we propose a novel framework called Twin
Support Vector Regression with Privileged Information (TSVR+), which
aims to improve the predictive performance of twin support vector regression by incorporating additional privileged information during the learning process. This TSVR+ introduces a twin-learning strategy that utilizes both original and privileged features to construct two non-parallel
boundary functions. One for positive deviations and the other for negative deviations and each boundary function is associated with a different
set of support vectors. Then, by solving two smaller quadratic programming problems, this proposed method achieves faster learning and better
prediction performance. Notably, this twin-boundary approach equipped
with privileged information provides a more robust representation of the
relationship between input features and target values, allowing for better
modeling of complex regression problems. Experimental results on various datasets demonstrate the effectiveness of our approach compared to
other regression methods.

Keywords: Twin support vector regression · Twin-learning strategy ·
Privileged information

1 Introduction

Support Vector Regression (SVR) [1] is a machine learning algorithm that has
been widely applied in various domains, including, finance, engineering, and
healthcare, where accurate regression is crucial. Generally, SVR aims to find
a function that can approximate the relationship between input features and
their corresponding target values. This is achieved by defining a hyperplane that
maximizes the margin around the training data points, while allowing a certain
degree of tolerance for errors or deviations from the actual targets. Recently,
a lot of SVR based method have been proposed in kinds of fields and shown
promising performance [2–4,6,7,10,14,15]. Among them, Least Square Support

© The Author(s), under exclusive license to Springer Nature Singapore Pte Ltd. 2023
E. Chen et al. (Eds.): BigData 2023, CCIS 2005, pp. 125–135, 2023.
https://doi.org/10.1007/978-981-99-8979-9_10

Vector Regression (LS-SVR) [10] adopts the quadratic loss function to measure the empirical risk along with the regularization term. Twin Support Vector Regression (TSVR) [3,6] extends the traditional SVR, aiming at generating two functions such that each one determines the down- or up-bound of the unknown regressor. TSVR allows for better interpretability by providing separate sets of support vectors for each target variable. Shao et al. proposed Least Squares Twin Bounded Support Vector Regression (TBSVR) [9] to reduce computational complexity. Moreover, an ensemble support vector regression model is proposed to establish the relationship between health indicator and battery state-of-health [2]. Considering that some existing methods fail to consider correlations between outputs or suffer from high computational complexity and sensitivity to parameters due to noise in multi-output regression tasks, Li et al. [4] proposed a method called multi-output twin support vector regression (M-TSVR) to overcome these issues. While TSVR is a powerful algorithm, there are few works to fully exploit the additional information to deal with the problems under some situations with privileged information, which causes the performance degradation. The concept of privileged information was introduced by Vapnik et al. [11] which can be described as any additional relevant data that captures valuable insights, such as expert annotations, additional measurements, or other correlated attributes. Utilizing privileged information can be beneficial in various domains and scenarios, especially when it provides additional knowledge, data, or insights that can improve the model's learning and predictive capabilities, leading to better performance on unseen data during testing or deployment. Notably, privileged information based models aim to leverage additional information that is not available during the testing or prediction phase [5,12,13]. Related to regression tasks, based on SVR, Shu et al. [8] presented a unified framework, called V-SVR+, that systematically addresses three forms of privileged information: continuous, ordinal, and binary by integrating these types of privileged information into the learning process of SVR using three distinct losses.

To our knowledge, how to effectively utilize of privileged information is still an open issue in the field. In this paper, we aim to achieve the model of TSVR+ by using privileged information paradigms. In TSVR+, the objective is to leverage privileged information during the training phase to develop a model that imposes additional constraints on the solution in the original space. Specifically, we define two linear correcting (slack) function in the privileged information space to estimate slack variables in the standard twin SVR method using privileged information. Thus, two separate sets of support vectors from original data and privileged information are constructed for each target variable, capturing both the positive and negative correlations with the input features. These support vectors form the basis for modeling the relationships between the input features and the target variables in a twin-like manner. The optimization problem in TSVR+ involves minimizing a joint objective function that incorporates both the individual target variable errors and the twin relationship constraints. This ensures that the predicted values not only fit the training data well but also maintain coherence among the target variables. In terms of computational efficiency, TSVR+ benefits from the sparsity of the solution obtained through the support vector framework. This enables faster training and prediction time compared to other regression methods.

The rest of this paper is shown as follows. In Sect. 2, we introduce some preliminaries related to our work. Then, in Sect. 3, we propose the twin support vector regression with privileged information. In Sect. 4, extensive experiments are conducted to evaluate the effectiveness of our method. Finally, we conclude our work in Sect. 5.

2   Related Works

2.1   Support Vector Regression

The SVR [1] would like to find a linear regression function $f(x) = w^T x + b$, tolerating a small error in fitting this given dataset (A, Y), where $w \in R^n$ and $b \in R$. This can be achieved by utilizing the ϵ–insensitive loss function that sets an ϵ–insensitive tube around the data, within which errors are discarded. Also, applying the idea of SVM, the function $f(x)$ is made as flat as possible in fitting the training dataset. The SVR can be formulated as the following constrained minimization problem:

$$
\begin{aligned}
\min_{w,b} \quad & \frac{1}{2}\|w\|^2 + C(e^T \xi + e^T \xi^*) \\
s.t. \quad & Y - (Aw + eb) \geq e\epsilon + \xi, \xi \geq 0, \\
& (Aw + eb) - Y \geq e\epsilon + \xi^*, \xi^* \geq 0,
\end{aligned}
\tag{1}
$$

where $C > 0$ is the regularization factor that weights the tradeoff between the fitting errors and the flatness of the linear regression function, ξ and ξ^* are the slack vectors reflecting if the samples locate into the ϵ–insensitive tube or not, e is the vector of ones of appropriate dimensions.

2.2   Twin Support Vector Regression

In this section, we introduce the Twin Support Vector Regression (TSVR). TSVR is similar to Twin Support Vector Machine (TSVM) in spirit, as it also derives a pair of nonparallel planes around the data points. However, there are some differences in essence. First, the targets of TSVR and TSVM are different, TSVR aims to find the suitable regressor while TSVM is to construct the classifier. Second, each of the two QPPs in the TSVM pair has the formulation of a typical SVM, except that not all patterns appear in the constraints of either problem at the same time, while all data points appear in the constraints of each of the two QPPs in the TSVR pair. Third, the TSVM pair finds two hyperplanes such that each plane is closer to one of the two classes and is as far as possible from the other, whereas the TSVR pair finds the ϵ–insensitive up-bound and down-bound functions for the end regressor. TSVR is obtained by solving the following pair of QPPs:

$$\min_{w_1, b_1, \xi} \frac{1}{2} \|Y - e\epsilon_1 - (Aw_1 + eb_1)\|^2 + c_1 e^T \xi \tag{2}$$
$$s.t. \quad Y - (Aw_1 + eb_1) \geq e\epsilon - \xi, \ \xi \geq 0,$$

$$\min_{w_2, b_2, \eta} \frac{1}{2} \|Y + e\epsilon_2 - (Aw_2 + eb_2)\|^2 + c_2 e^T \eta \tag{3}$$
$$s.t. \quad (Aw_2 + eb_2) - Y \geq e\epsilon_2 - \eta, \ \eta \geq 0,$$

where $c_1, c_2 > 0$, $\epsilon_1, \epsilon \geq 0$ are parameters, and ξ, η are slack vectors. The TSVR algorithm finds two functions $f_1(x) = w_1^T x + b_1$ and $f_2(x) = w_2^T x + b_2$, each one determines the ϵ–insensitive down- or up-bound regressor.

3   Twin Support Vector Regression with Privileged Information

In this section, we introduce an efficient approach to TSVR which we have termed as Twin Support Vector Regression with privileged information (TSVR+). In the Learning Using Privileged Information (LUPI) setup, during the training phase, instead of tuples of features and labels, we are given triplets $(x_i, x_i^*, y_i)_{i=1}^{N}$, where feature vectors x^* represent the additional (i.e., privileged) information. During the testing phase, features from the privileged space X^* are not available. The goal of LUPI is to exploit the privileged information during the training phase to learn a model that further constrains the solution in the original space X, and thus it can more accurately describe the testing data. Given the training data points (A, A^*, Y), where each row of A and A^* represent a training point and an additional information representation respectively, unlike TSVR, in this paradigm, the slack variables ξ and η are parameterized as a function of privileged information, i.e., $\xi = A^* w_1^* + e^* b_1^*$ and $\eta = A^* w_2^* + e^* b_2^*$. Thus, the TSVR+ algorithm, which implements LUPI in the training phase, solves the following minimization problem:

$$\min_{w_1, b_1, w_1^*, b_1^*} \frac{1}{2} \|Y - e\epsilon_1 - (Aw_1 + eb_1) - (A^* w_1^* + e^* b_1^*)\|^2 + c_1 e^T (A^* w_1^* + e^* b_1^*)$$
$$s.t. \quad Y - (Aw_1 + eb_1) \geq e\epsilon - (A^* w_1^* + e^* b_1^*), \ (A^* w_1^* + e^* b_1^*) \geq 0, \tag{4}$$

$$\min_{w_2, b_2, w_2^*, b_2^*} \frac{1}{2} \|Y + e\epsilon_2 - (Aw_2 + eb_2) - (A^* w_2^* + e^* b_2^*)\|^2 + c_2 e^T (A^* w_2^* + e^* b_2^*)$$
$$s.t. \quad (Aw_2 + eb_2) - Y \geq e\epsilon_2 - (A^* w_2^* + e^* b_2^*), \ (A^* w_2^* + e^* b_2^*) \geq 0, \tag{5}$$

where $c_1, c_2 > 0, \epsilon_1, \epsilon_2 > 0$ are parameters. In Eq. (4) and Eq. (5), two non-parallel boundary functions are introduced by introducing a twin-learning strategy that utilizes both original and privileged features. Specifically, the function $f_1(x)$ determines the ϵ_1–insensitive down-bound regressor, while the function $f_2(x)$ determines the ϵ_2–insensitive up-bound regressor. The first term in the objective function of (4) or (5) is the sum of squared distances based on both original and privileged features. As we can obtain, TSVR+ is comprised of a pair of QPPs such that each QPP determines the one of up- or down-bound function by using only one group of constraints. Hence, TSVR+ gives rise to two smaller sized QPPs. To derive the dual QPPs of TSVR+, we first introduce the Lagrangian function for the problem (4):

$$
\begin{aligned}
&L(w_1, b_1, w_1^*, b_1^*, \alpha, \beta) \\
&= \frac{1}{2}\|Y - e\epsilon_1 - (Aw_1 + eb_1) - (A^*w_1^* + e^*b_1^*)\|^2 + c_1 e^T(A^*w_1^* + e^*b_1^*) \quad (6) \\
&\quad - \alpha^T(Y - (Aw_1 + eb_1) - e\epsilon + (A^*w_1^* + e^*b_1^*)) - \beta^T(A^*w_1^* + e^*b_1^*),
\end{aligned}
$$

where $\alpha = (\alpha_1; \alpha_2; \ldots; \alpha_l)$ and $\beta = (\beta_1; \beta_2; \ldots; \beta_l)$ are the Lagrangian multiplier vectors. The Karush Kuhn Tucker (KKT) optimality conditions for the problem Eq. (6) are given by:

$$
\begin{aligned}
&- A^T(Y - e\epsilon_1 - (Aw_1 + eb_1) - (A^*w_1^* + e^*b_1^*)) + A^T\alpha = 0, \\
&- e^T(Y - e\epsilon_1 - (Aw_1 + eb_1) - (A^*w_1^* + e^*b_1^*)) + e^T\alpha = 0, \\
&Y - (Aw_1 + eb_1) \geq e\epsilon - (A^*w_1^* + e^*b_1^*), \ (A^*w_1^* + e^*b_1^*) \geq 0, \\
&\alpha^T(Y - (Aw_1 + eb_1) - e\epsilon + (A^*w_1^* + e^*b_1^*)) = 0, \ \alpha \geq 0, \quad (7) \\
&\beta^T(A^*w_1^* + e^*b_1^*) = 0, \ \beta \geq 0, \\
&c_1 e - \alpha - \beta = 0.
\end{aligned}
$$

Since $\beta \geq 0$, we have $0 \leq \alpha \leq c_1 e$. By calculating Eq. (7), it can be obtained:

$$
- \begin{bmatrix} A^T \\ e^T \end{bmatrix}((Y - e\epsilon_1) - [A \ e]\begin{bmatrix} w_1 \\ b_1 \end{bmatrix} - [A^* \ e^*]\begin{bmatrix} w_1^* \\ b_1^* \end{bmatrix}) + \begin{bmatrix} A^T \\ e^T \end{bmatrix}\alpha = 0. \quad (8)
$$

Define

$$
G = [A \ e], \ f = Y - e\epsilon_1, \ u_1 = \begin{bmatrix} w_1 \\ b_1 \end{bmatrix} \ G^* = [A^* \ e^*], \ u_1^* = \begin{bmatrix} w_1^* \\ b_1^* \end{bmatrix}, \quad (9)
$$

then we have

$$
-G^T f + G^T G u_1 + G^T G^* u_1^* + G^T \alpha = 0, \quad (10)
$$

i.e.,

$$
-G^T f + G^T G u_1 + G^T G^* u_1^* + G^T \alpha = 0,
$$
$$
i.e., \quad u_1 = (G^T G)^{-1} G^T(f - \alpha) - (G^T G)^{-1} G^T G^* u_1^*. \quad (11)
$$

Notice that $G^T G$ is always positive semidefinite, it is possible that it may not be well conditioned in some situations. To overcome this ill-conditioning case, we introduce a regularization term σI, where σ is a very small positive number. Therefore, problem (11) is modified to

$$
u_1 = (G^T G + \sigma I)^{-1} G^T(f - \alpha) - (G^T G + \sigma I)^{-1} G^T G^* u_1^*. \quad (12)
$$

Substituting Eq. (12) and the above KKT conditions into (9) and discarding the all constant terms, we obtain the dual QPP for Eq. (7) as follows:

$$\min_{\alpha} \frac{1}{2}\alpha^T G(G^T G)^{-1}G^T \alpha - f^T G(G^T G)^{-1}G^T \alpha + (f - G^* u_1^*)^T \alpha \tag{13}$$

$$s.t. \quad 0 \le \alpha \le c_1 e.$$

Similarly, we consider the problem (5) and obtain its dual as

$$\min_{\alpha} \frac{1}{2}\gamma^T G(G^T G)^{-1}G^T \gamma + h^T G(G^T G)^{-1}G^T \gamma - (h - G^* u_2^*)^T \gamma \tag{14}$$

$$s.t. \quad 0 \le \gamma \le c_2 e,$$

where $h = Y + e\epsilon_2$, $u_2^* = \begin{bmatrix} w_2^* \\ b_2^* \end{bmatrix}$ and further

$$u_2 = \begin{bmatrix} w_2 \\ b_2 \end{bmatrix} = (G^T G)^{-1}G^T (h + \gamma) - (G^T G)^{-1}G^T G^* u_2^*. \tag{15}$$

Note that in the dual QPPs (13) and (14), we have to compute the inversion of matrix $G^T G$ of size $(n + 1) \times (n + 1)$. In general, n is much smaller than the number of training samples. Further, comparing the above two QPPs with the dual QPP (4) of standard SVR, we find the latter has another equality constraint, indicating TSVR+ is far faster in comparison to the standard SVR in order to find the optimal solution. Once the vectors u_1 and u_2 are known from (12) and (15), the two up- and down-bound functions are obtained. Then the estimated regressor is constructed as follows

$$f(x) = \frac{1}{2}(f_1(x) + f_2(x)) = \frac{1}{2}(w_1 + w_2)^T x + \frac{1}{2}(b_1 + b_2). \tag{16}$$

4   Experiment

4.1   Datasets and Setting

In this section, to evaluate the effectiveness of the proposed TSVR+, we conduct experiments on three benchmark datasets[1] and three synthetic datasets. Specifically, the benchmark datasets includes Wisconsin Breast Cancer dataset, Boston housing dataset and Marriage-and-Divorce datasets. Wisconsin Breast Cancer (WBC) dataset was obtained from the University of Wisconsin Hospitals. It has 683 instances and each has 10 features. Boston housing dataset (BH) contains a total of 506 samples. Each sample includes 13 feature information and actual housing prices. Marriage-and-Divorce (MD) dataset contains 100 samples with 31 features. The first 30 columns are features (inputs). The 31th column is Divorce Probability (Target). Besides, three synthetic datasets consist of 1000 data samples that are generated randomly using the following regression models:

---

[1] https://www.kaggle.com/datasets.

Synthetic data1 (SynD1):

$$f(x) = 0.25[1.5(1-x_1) + e^{2x_1-1}\sin(3\pi(x_1-0.6)^2) \\ + e^{3(x_2-0.5)}\sin(4\pi(x_2-0.9)^2)], \tag{17}$$

Synthetic data2 (SynD2):

$$f_2(x) = \sin(2\pi x_1) + 4(x_2 - 0.5)^2, \tag{18}$$

Synthetic data2 (SynD3):

$$f_3(x) = 20 - 20e^{-0.2\sqrt{x_1/3}} - e^{\cos(2\pi x_1)} + e \\ + x_2^2 - 10\cos(2\pi x_2) + 10, \tag{19}$$

where $x = (x_1, \cdots, x_n)^T \in [0,1]^n$, we use $n = 2$.

To further demonstrate the performance of our proposed TSVR+, we compared with several SVR based methods, including, Support Vector Regression (SVR), Twin Support Vector Regression (TSVR), Least Square Twin Support Vector Regression (LS-TSVR), Twin Parametric Insensitive Support Vector Regression (TPISVR), Twin Bounded Support Vector Regression (TBSVR). We adopt the following experimental setting: for all the methods, the percentage of testing data is selected as 30%. Thus, we obtain 70% data samples as the gallery set. Motivated by the work of the Mechanical Turk image annotation [8], samples can be defined to achieve the privileged information, we adopt a simple selection strategy to label half of the data samples among the gallery set as privileged information. These privileged samples are only used to train the model of TSVR+. The parameters for all methods are selected from the set $\{10^i | i = -10, \cdots, 10\}$. Without loss generality, denote m as the number of testing samples, $\widehat{y}_i$ as the prediction value of y_i, and $\overline{y} = \frac{1}{m}\sum_i y_i$ as the average value of $y_1, \cdots, y_m$. Then, we use the following five criterions for algorithm evaluation:

(1) Mean absolute error (MAE): $\frac{1}{N}\sum_{i=1}^{N}\frac{\|y_i-\widehat{y}_i\|}{\|y_i+\widehat{y}_i\|}$.

(2) Root mean square error (RMSE): $\sqrt{\frac{1}{N}\sum_{i=1}^{N}\|y_i - \widehat{y}_i\|}$.

(3) Sum squared error of testing (SSE): $\sum_{i=1}^{m}(y_i - \widehat{y}_i)^2$.

(4) Ratio between sum squared error and sum squared deviation of testing samples (S1): $\sum_{i=1}^{m}(y_i - \widehat{y}_i)^2 / \sum_{i=1}^{m}(y_i - \overline{y})^2$.

(5) Ratio between interpretable sum squared deviation and real sum squared deviation of testing samples (S2): $\sum_{i=1}^{m}(\widehat{y}_i - \overline{y})^2 / \sum_{i=1}^{m}(y_i - \overline{y})^2$.

For each evaluation criterion, the smaller it is, the better regressing performance of learning method has. The results of all methods on three benchmark datasets and three synthetic datasets are listed in Table 1 and Table 2.

Table 1. Performance of all methods on different datasets

Dataset	Metric	SVR	TSVR	LS-TSVR	TBSVR	TRISVR	TSVR+
WBC	MAE	0.0857	0.0705	0.0808	0.0748	0.1212	**0.0401**
	RMSE	0.1154	0.1048	0.1187	0.1088	0.1339	**0.0847**
	SSE	2.7290	2.2497	2.8894	2.4265	3.6741	**1.4708**
	S1	2.2308	1.8391	2.3619	1.9836	3.0034	**1.2024**
	S2	1.2308	0.8405	1.3595	0.9569	2.0034	**0.2042**
BH	MAE	0.1161	0.1043	0.1242	0.1075	0.1086	**0.0824**
	RMSE	0.1825	0.1721	0.1877	0.1743	0.1760	**0.1594**
	SSE	3.3968	3.0220	3.5919	3.0973	3.1598	**2.5903**
	S1	1.6804	1.4949	1.7769	1.5322	1.5631	**1.2814**
	S2	0.6804	0.5288	0.7494	0.5511	0.5769	**0.2931**
MD	MAE	1.2707	1.0094	1.0492	0.9760	1.0896	**0.8658**
	RMSE	1.3042	1.1812	1.3241	1.1932	1.1568	**1.0496**
	SSE	51.0278	41.8572	52.5943	42.7083	40.1484	**33.0503**
	S1	19.7124	16.1697	20.3176	16.4985	15.5096	**12.7676**
	S2	18.7124	16.9029	21.4308	17.1904	14.3726	**13.5949**

Table 2. Performance of all methods on synthetic datasets

Dataset	Metric	SVR	TSVR	LS-TSVR	TBSVR	TRISVR	TSVR+
SynD1	MAE	0.4104	0.4974	0.4113	0.4711	0.7652	**0.2138**
	RMSE	0.4450	0.5259	0.4454	0.5642	0.9022	**0.2497**
	SSE	59.4092	82.9615	59.5086	95.4885	244.1820	**18.7102**
	S1	6.6888	9.3406	6.7000	10.7510	27.4923	**3.3397**
	S2	5.6888	8.3862	5.7276	9.7297	27.8560	**2.4361**
SynD2	MAE	0.4433	0.2361	0.4616	0.4515	0.3904	**0.2125**
	RMSE	0.5032	0.2795	0.5134	0.5156	0.4583	**0.2622**
	SSE	75.9525	23.4334	79.0862	79.7511	63.0068	**20.6229**
	S1	4.4657	1.3778	4.6499	4.6890	3.7045	**1.1676**
	S2	3.4657	1.3176	4.4173	5.4594	1.5812	**1.0620**
SynD3	MAE	0.5111	0.4766	0.5101	0.5109	0.3750	**0.3595**
	RMSE	0.6113	0.5806	0.6086	0.7316	**0.5260**	0.5960
	SSE	112.1070	101.1247	111.1260	160.5906	82.9967	**38.7813**
	S1	3.3230	2.9974	3.2939	4.7601	2.4601	**1.1495**
	S2	2.3230	2.0224	2.3185	3.4995	1.5664	**0.1721**

Table 3. Time analysis on different datasets (s)

Method	SVR	TSVR	LS-TSVR	TBSVR	TRISVR	TSVR+
WBC	0.2917	0.0522	0.0264	0.0692	0.0524	0.0458
BH	0.3440	0.0458	0.0192	0.0585	0.0789	0.0726
MD	0.8552	0.2689	0.3555	0.4183	0.4420	0.3212

4.2 Experiments Analysis

From the results in Table 1 we can obtain that our proposed method achieves competitive performance compared with the existing methods under all criterions. The predictive performance of TSVR+ is smaller than those of SVR and other TSVR based regression methods. Specifically, on WBC, the achieved results of MAE for TSVR, LS-TSVR, TBSVR, TRISVR and TSVR+ are 0.0705, 0.0808, 0.0748, 0.1212, and 0.0401 respectively, the S1 for TSVR, LS-TSVR, TBSVR, TRISVR, and TSVR+ are 1.8391, 2.3619, 1.9836, 3.0034, and 1.2024, which indicates that TSVR, TBSVR, and our TSVR+ with small values mean good agreement between estimations and real-values. Compared with TSVR based models, we can find that SVR leads to the worst predictive performance, which demonstrates that twin support vector learning mechanism with a pair of nonparallel up- and down-bound functions is superior to support vector learning. Moreover, from Table 2, on SynD1 and SynD2 datasets, we can observe that our method exhibits the smallest predictive results under all evaluation indicators. On SynD3, the performance also shows better under MAE, RMSE, S1 and S2 for TSVR+ compared with other methods. Therefore, from the experimental results on benchmark and synthetic datasets, we can conclude that the proposed TSVR+ is an effective and competitive regressor for the reason that it can additionally utilize the privileged information to train a robust model. Besides, to show a more intuitive regression analysis, we show the true value and the prediction value of all methods on synD2 dataset in Fig. 1(a)–(f). Obviously, our method obtains better performance.

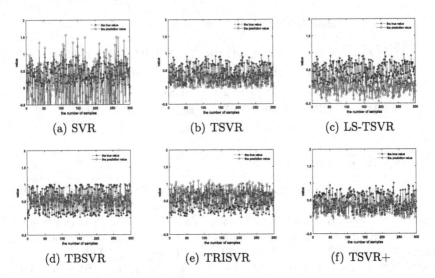

(a) SVR   (b) TSVR   (c) LS-TSVR

(d) TBSVR   (e) TRISVR   (f) TSVR+

Fig. 1. Regression analysis on synD2 dataset

4.3   Computing Time

In this section, we provide the time analysis (s) of all methods on three bench-mark datasets in Table 3. In general, the time complexity of regression algorithms depends on the number of training samples and the dimensionality of the feature space. Training a SVR model typically involves solving a quadratic optimization problem, which can be computationally intensive (with 0.2917, 0.3440 and 0.8552 on WBC, BH and MD, respectively), while training TSVR based methods with two small QPP problems result in a faster speed. For TSVM+, it modifies existing TSVR to incorporate privileged information, which may require additional computations. For example, TSVR and LS-TSVR show fast learning results. Notably, our method can achieve better computational efficiency than TBSVR and TRISVR on most cases.

5   Conclusions

Considering the situations with more additional information in practical application, we aim to enhance the performance of TSVM with privileged information learning framework. Our method utilizes a twin-learning strategy that leverages both original and privileged features to create two non-parallel boundary functions. By solving smaller quadratic programming problems, our approach facilitates faster learning and enhances prediction performance. Experimental evaluations conducted on various datasets demonstrate the effectiveness of our approach when compared to other existing methods.

Acknowledgment. This work is supported by the National Natural Science Foundation of China under Grant No. (62106107), and by the Natural Science Foundation of

Jiangsu Province, China (Youth Fund Project) under Grant No. (BK20210560) and by the Natural Science Research of Jiangsu Higher Education Institutions of China under Grant No. (21KJB520011), and by the Start Foundation of Nanjing University of Posts and Telecommunications under Grant No. (NY222024).

References

1. Drucker, H., et al.: Support vector regression machines, pp. 779–784 (1997)
2. Guo, Y., Huang, K., Yu, X., Wang, Y.: State-of-health estimation for lithium-ion batteries based on historical dependency of charging data and ensemble SVR. Electrochim. Acta **428**, 140940 (2022)
3. Huang, H., Wei, X., Zhou, Y.: An overview on twin support vector regression. Neurocomputing **490**, 80–92 (2022)
4. Li, Y., Sun, H., Yan, W., Zhang, X.: Multi-output parameter-insensitive kernel twin SVR model. Neural Netw. **121**, 276–293 (2020)
5. Liu, B., Liu, L., Xiao, Y., Liu, C., Chen, X., Li, W.: Adaboost-based transfer learning with privileged information. Inf. Sci. **593**, 216–232 (2022)
6. Peng, X.: TSVR: an efficient twin support vector machine for regression. Neural Netw. **23**(3), 365–372 (2010)
7. Peng, X.: Efficient twin parametric insensitive support vector regression model. Neurocomputing **79**, 26–38 (2013)
8. Qi, Z., Tian, Y., Shi, Y.: A new classification model using privileged information and its application. Neurocomputing **129**, 146–152 (2014)
9. Shao, Y.H., Zhang, C.H., Yang, Z.M., Jing, L.: An ε-twin support vector machine for regression. Neural Comput. Appl. **23**(1), 175–185 (2013)
10. Suykens, J.A.K., Vandewalle, J.: Least squares support vector machine classifiers. Neural Process. Lett. **9**(3), 293–300 (1999)
11. Vapnik, V., Vashist, A.: A new learning paradigm: learning using privileged information. Neural Netw. **22**, 544–557 (2009)
12. Vapnik, V., Izmailov, R.: Learning using privileged information: similarity control and knowledge transfer. JMLR.org (2015)
13. Wu, Z., et al.: LR-SVM+: learning using privileged information with noisy labels. IEEE Trans. Multimed. **24**, 1080–1092 (2022)
14. You, J.: Robust online support vector regression with truncated ε-insensitive pinball loss. Mathematics **11**, 709 (2023)
15. Zhang, Z., Hong, W.C.: Application of variational mode decomposition and chaotic grey wolf optimizer with support vector regression for forecasting electric loads. Knowl.-Based Syst. **228**, 107297 (2021)

Detecting Social Robots Based on Multi-view Graph Transformer

Mingyuan Li[1,2,3,4(✉)], Zhonglin Ye[1,2,3,4], Haixing Zhao[1,2,3,4], Yuzhi Xiao[1,2,3,4],
and Shujuan Cao[1,2,3,4]

[1] College of Computer, Qinghai Normal University, Xining, China
Mingyuan1007_li@foxmail.com
[2] The State Key Laboratory of Tibetan Intelligent Information Processing and Application,
Xining, China
[3] Tibetan Information Processing and Machine Translation Key Laboratory of Qinghai
Province, Xining, China
[4] Key Laboratory of Tibetan Information Processing, Ministry of Education, Xining, China

Abstract. With the development of social media, social robots are increasingly interfering in political elections and economic issues, and detecting social robots has become a long-standing but unresolved problem. Traditional machine learning models or methods have gradually become ineffective with the evolution of social robots. In contrast, analyzing the social network structure where social robots are located has become a more effective method. However, at present, most of these methods are based on a single view, ignoring the importance of multi-view information in social networks. At the same time, existing methods have not considered the influence of potential topic structures. Therefore, to solve the above problems, this paper proposes a new framework based on multi-view (Multi-View Graph Transformer, MV-GT). Specifically, MV-GT includes a topic module, a graph enhancement module, a multi-view Graph Transformer module, and a multi-view attention module, which can explore the complementarity, consistency, and semantic relevance of multiple different views in online social networks. Experimental results show that MV-GT outperforms many existing methods and also demonstrates the effectiveness of multi-view and topic structures in detecting social robots.

Keywords: Multi-View · Graph Transformer · Topic Structures · Detecting Social Robots

1 Introduction

Twitter, as a social platform with global influence and coverage, allows users to express their opinions. However, it also contains a large number of social robots that disguise themselves as normal users and participate in public discussions while intervening in the dissemination process [1, 2], and thus influencing public sentiment [3]. As a result, scholars have been devoted to combating robots in the past decade.

© The Author(s), under exclusive license to Springer Nature Singapore Pte Ltd. 2023
E. Chen et al. (Eds.): BigData 2023, CCIS 2005, pp. 136–148, 2023.
https://doi.org/10.1007/978-981-99-8979-9_11

Early research on social robot detection heavily relied on feature engineering. Specifically, researchers defined and extracted unique features of social robots from raw data through domain knowledge, and applied them to machine learning models [4–6]. In order to further explore the differentiation between normal users and social robots, researchers proposed the CNN-LSTM model to analyze the internal and external factors of social behavior in detail [7]. With the rise of graph neural networks, researchers began to use graphs to describe the social association structure between social robots and normal users [8]. Graph mining algorithms have also been gradually applied to the detection of social robots [9]. In this process, researchers have explored from homogeneous graphs to heterogeneous graphs, with some focusing on heterogeneous networks [10] and multi-relationship networks [21] to explore the subtle differences between social robots and normal users. Although these methods have achieved some success, they have several obvious problems. For example, the social correlation among users' behaviors under the same topic in social networks has not been taken into account; the complex behavior information of social robots has not been captured from a multi-view perspective.

To address the above issues, a new topic structure is proposed to understand the influence of other users within the community on users' behavior. Then, this paper proposes a new Twitter bot detection framework to analyze the significant differences between users and social bots in social networks from a multi-view perspective. The main contributions of this paper are summarized as follows:

(1) This paper introduces a new topic structure in the heterogeneous information network, which can help better understand users' behavior on social media and how they are influenced by other users, while also helping to identify subtle differences between bots and real users.
(2) This paper proposes a novel end-to-end Twitter bot detection framework, which can use multi-view Graph Transformer for fine-grained analysis of different views in social networks, and dynamically learn the importance and correlation between different views through attention mechanism, thus improving the accuracy of identifying and detecting Twitter bot accounts.
(3) Experiments conducted on publicly available datasets show that the proposed model outperforms state-of-the-art methods in terms of performance. Further analysis also confirms the effectiveness of the proposed topic structure and multi-view model.

2 Related Work

Userinfo-Base Methods. Early research mainly relied on manual feature extraction and combining machine learning models to differentiate between normal users and social bots [4, 11]. In order to conduct a more in-depth analysis, researchers introduced features such as time [5, 6] and text semantics [12, 13] based on this foundation. Compared with early feature extraction methods and machine learning models, these deep learning models extract higher-level features from user information and behavior through multi-layer neural networks [14, 15]. These rich features not only help researchers effectively differentiate between normal users and social bots, but can also capture more complex features, such as joint features [16], user interaction patterns [17, 18], and contextual features [19], making the social bot identification process more accurate and reliable.

Graph-Based Methods. Recent advances in graph neural networks have helped to better understand the underlying relationships between normal users and social bots, improving detection efficiency. [8] presented the detection process as a node classification problem in a graph, and for the first time, applied graph convolutional networks to social bot detection. [20] considered the different types of relationships between users in social networks and identified social bots by constructing a heterogeneous information network. To further analyze this, [21] added influence heterogeneity and relationship heterogeneity to the heterogeneous information network to improve the model's performance. Meanwhile, [22] considered the similarity of robot behavior in different social networks and added neighborhood awareness to a multi-relationship network to explore differences in robot behavior in multiple domains. In recent work, Reference [10] combined reinforcement learning with GNN to adaptively determine the most appropriate multi-hop neighborhood and layer in the GNN architecture, thereby improving the model's detection performance for social bots.

Multi-view Methods. Multi-view learning refers to the use of complementary information from multiple features or patterns to improve performance, and has achieved significant success in many fields. [23] proposed the Multi-View Graph Convolutional Network (MV-GCN) model, which integrates GCN into multi-view learning by utilizing the interaction relationships and content information between different node objects. [24] built upon this model and proposed MGAT, which aggregates node representations of each view through an attention mechanism.

In this paper, inspired by the aforementioned research, we propose a multi-view based framework for social bot detection, which adaptively learns node representations for each view and integrates information from various views into a unified node feature representation. Additionally, it allows for interaction between different views to better represent the learned content.

3 Methodology

Figure 1 illustrates the framework structure of MV-GT. Specifically, MV-GT includes a topic structure module for constructing latent semantic structures, a graph augmentation module for extending multiple views to multi-channel and multi-view, a multi-view Graph Transformer module, and a multi-view interaction attention module for social bot detection.

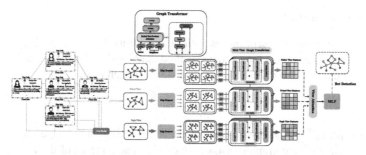

Fig. 1. Framework Architecture Diagram.

3.1 Topic Graph Construction

Due to the existence of a large number of short texts in users' tweets, the sparsity of these texts can lead to poor co-occurrence effects, which can severely impact the establishment of topic models. In this paper, we construct pseudo-tweets by concatenating all tweets from the same user, as follows,

$$d_i = \|_{j=1}^{J} d_i^j, \tag{1}$$

where $\|$ Indicates splicing operation, d_j^i, d_i denote the i-th tweet of user j and the constructed pseudo-tweet. LDA uses a Dirichlet prior with fixed number of topics to represent the probability distribution over m topics,

$$D_i = (P(T_1|d_i), P(T_2|d_i), P(T_3|d_i), \ldots, P(T_m|d_i)), \tag{2}$$

where D_i denotes the distribution of tweet topics for user i, The set T represents the topics that appear in n user tweets, $P(T_j|d_i)$ denotes the probability that user i tweet d_i belongs to topic T_i. Find users under each topic by filtering the largest T_i,

$$T = \{T_1, T_2, T_3, \ldots, T_m\}, \tag{3}$$

$$T_i = \{u_1, u_2, u_3, \ldots, u_s\}, \tag{4}$$

where T represents the topic cluster, T_i denotes the s-th topic cluster, m denotes the number of topics, i denotes the users under the i-th topic cluster, and s denotes the number of users under the corresponding topic cluster; moreover, the number of users under each topic cluster is not necessarily the same. To further reduce the denseness of the topic network, the users under each topic cluster are concatenated with probability p to obtain the final topic structure network. Finally, the users are represented as x_i and are transformed into the initial features of the GNN by a fully connected layer,

$$x_i^{(0)} = \sigma(W_1 \cdot x_i + b_1). \tag{5}$$

3.2 Graph Augmentation

To reduce the effects of overfitting as well as over-smoothing, DropEdege is chosen as the graph structure data enhancement method, and by DropEdege on each view, the multi-view is extended to multi-view multi-channel, which is represented as follows,

$$S_i(G_z) = (V, M_i \odot E_z). \tag{6}$$

where $M_i \in \{0, 1\}^{|V|}$ is the mask vector of channel i acting on the edge set E_v of view z and $S_i(G_z)$ denotes the graph of view z channel i. The patterns of the local structure of the nodes under each view are captured by multiple channels, multiple subgraphs, further making the node representations more robust and noise-resistant.

3.3 Mult-view Graph Transformer

In the Multi View Graph Transformer module, the multi head attention mechanism is introduced into graph learning for multiple views, taking into account the specific features of each view. Specifically, the c-th attention head mechanism of channel s and node i under view z is computed by the node features initialized above,

$$q_{c,i}^{z,s(l)} = w_{c,q}^{z,s(l)} \cdot x_i^{(l-1)} + b_{c,q}^{z,s(l)}, \tag{7}$$

$$k_{c,j}^{z,s(l)} = w_{c,k}^{z,s(l)} \cdot x_j^{(l-1)} + b_{c,k}^{z,s(l)}, \tag{8}$$

$$v_{c,i}^{z,s(l)} = w_{c,q}^{z,s(l)} \cdot x_i^{(l-1)} + b_{c,v}^{z,s(l)}, \tag{9}$$

where q, k, and v denote $query, key$, and $value$ in the attention mechanism, and (l) denotes the l-th layer of the model. $w_{c,q}^{z,s(l)}, w_{c,k}^{z,s(l)}, b_{c,q}^{z,s(l)}, b_{c,k}^{z,s(l)}$ denote the parameters that can be learned by the attention head c of channel s under view z.Inspired by [25], the features $e_{i,j}$ of the edges of the different attention heads of the channels s under view z are also added to the vector calculation as supplementary information for each layer of the channels s under view z,

$$e_{c,i,j}^{z,s} = W_{c,e}^{z,s(l)} \cdot e_{i,j}^{z,s} + b_{c,e}^{z,s}, \tag{10}$$

further it can be modeled by the attention weights between different nodes,

$$\alpha_{c,i,j}^{z,s(l)} = \frac{q_{c,i}^{z,s(l)}, k_{c,j}^{z,s(l)} + e_{c,i,j}^{z,s}}{\sum_{u \in N(i)^{z,s}} q_{c,i}^{z,s(l)}, k_{c,u}^{z,s(l)} + e_{c,i,j}^{z,s}}, \tag{11}$$

where $\alpha_{c,i,j}^{(l)}$ denotes the c-th attention head weight of the l-th layer between node i and node j, $<q, k> = \exp\left(q^T k / \sqrt{d}\right)$ is the exponential dot product function, d denotes the hidden size of each attention head, $N(i)^Z$ denotes the neighborhood of node i of channel s under view z. Finally, the node i of channel s under view z is obtained by message aggregation by representing,

$$x_i^{Z,S(l)} = \frac{1}{C}\sum_{c=1}^{C}\left[\sum_{j\in N(i)^{Z,S}} \alpha_{c,i,j}^{Z,S(l)}\left(v_{c,i}^{Z,S(l)} + e_{c,i,j}^{Z,S}\right)\right], \tag{12}$$

where $x_i^{Z,S(l)}$ is the node representation of layer l of channel s under view z and C is the number of heads of the attention mechanism. After completing the learned representation of the nodes under each channel, the nodes of each channel under view z are subjected to the splicing operation, to which the connection operation is added due to the layer-by-layer decay of the node features as follows,

$$x_i^{Z,S} = \|_{l=1}^{L} x_i^{Z,S(l)} + x_i^{(0)}, \tag{13}$$

where $x_i^{Z,S}$ is the final representation of node i of channel s under view z. The fusion of multiple channels is accomplished through the connection of cumulative summation, and to ensure the smoothness of the node features, the nonlinear function and normalization operation are applied to the above results, and the final representation of node i under view z is represented through u_i^Z as follows,

$$\beta_i^Z = sigmoid\left(W_i^Z\left(x_i^{(0)} + \sum_{s=1}^{S} u_i^{Z,S}\right) + b_i^Z\right), \tag{14}$$

$$u_i^Z = LeakyReLU\left(LayerNorm\left(\left(x_i^{(0)} + \sum_{s=1}^{S} u_i^{Z,S}\right) \odot \beta_i^Z + x_i^{Z,S} \odot \left(1 - \beta_i^Z\right)\right)\right). \tag{15}$$

3.4 Mult-view Attention

In order to convert the input features into high-level output features by a shared linear transformation of the learnable weight vector α applied to each node, the coefficients computed by the attention mechanism can be expressed as,

$$\alpha_i^z = \frac{\exp\left(LeakyReLU\left(\alpha\left(W \cdot u_i^Z + b\right)\right)\right)}{\sum_{k=1}^{K} \exp\left(LeakyReLU\left(\alpha\left(W \cdot u_i^k + b\right)\right)\right)}, \tag{16}$$

where W and b are learnable parameters, α_i^z denotes the weight of different views, and further the final node can be represented as,

$$x_i = \sum_{z=1}^{Z} \alpha_i^Z \cdot u_i^Z, \tag{17}$$

where u_i^z denotes the representation of node i under view z. The global node representation x_i is generated by weighting and combining the nodes under different views with α_i^z as the weight parameter.

3.5 Training and Optimization

After the above-mentioned weighted fusion based on the feature weights of the different views through the attention mechanism, effectively combining the multi-view features, the final node representation will be obtained and will be used in the MLP layer as follows,

$$\hat{y}_i = softmax(W_2 \cdot LeakyReLU(W_1 \cdot x_i + b_1) + b_2), \tag{18}$$

where $\hat{y}_i$ is the predictive label of the model prediction, and W and b are both parameters that can be learned. The cross-entropy loss was then evaluated for all labeled data:

$$\text{Loss} = \sum_{i \in S} \left[-y_i \log \hat{y}_i - (1 - y_i) \log(1 - \hat{y}_i) \right] + \lambda \sum_{\omega \in \theta} \omega^2, \tag{19}$$

where s denotes the sets of users in the labeled data, y_i denotes the real label of the user, λ denotes the regular term coefficient, and θ denotes all learnable parameters.

4 Experiments

In this section, a large number of experiments are conducted to verify the effectiveness of the proposed method. First, the data set is presented, then the experimental results are shown, and finally the ablation experiments are used to further demonstrate the necessity of the framework composition.

4.1 Dataset

Since MV-GT is based on multiple views, datasets with specific graph structures need to be provided. In this paper, TwiBot-20 [26] was used for the main experiments. The details of this dataset are shown in Table 1.

Table 1. TwiBot-20 data set situation

Data item	Amount
User Node	229,573
User Tweet	33,488,192
User Attribute Item	8,723,736
User Relation(Edge)	33,716,171

TwiBot-20 covers a diverse range of bots and real users to better represent the real-world Twittersphere, so MV-GT can prove applicable to a variety of social bots. The same segmentation provided in the benchmark is followed, so the results are directly comparable to previous work.

4.2 Baselines

Our model was trained for 150 rounds, using AdamW as the optimizer, with a learning rate of 0.001, a weight decay of 0.0005, a number of topics of 9, and a concatenated edge probability of 0.4 for the topic view, all experiments were performed on an NVIDIA Tesla V100, Pytorch 1.12.1 was chosen as the deep learning framework.

This paper aims to provide a comprehensive comparison of various methods used for Twitter bot detection. In addition to some general methods, the following methods will be considered for this comparison: Cresci et al., [2], Lee et al., [5], Wei et al., [19], Yang et al., [28], Kudugunta et al. [30], Alhosseini et al., [8], SATAR [29], BotRGCN [20], RGT [21].

In order to better analyze the differences between each method, the performance of the model will be analyzed from several perspectives.

Table 2 shows the performance of the model evaluated in terms of the analysis of the underlying topic structure, the heterogeneous type of user relationship structure, the depth of the model, whether the model involves graph neural networks, and whether the model involves multi-view learning of multiple modalities.

Table 2. Multi-angle model analysis

Method	Theme	Heterogeneous	Deep	GNN-base	Mult-view	Accuracy	F1-score
Cresci et al.						0.4793	0.1072
Lee et al.						0.7456	0.7823
Miller et al.						0.4801	0.6266
Wei et al.			✓			0.7126	0.7515
Yang et al.						0.8191	0.8546
Kudugunta et al.			✓			0.8174	0.7515
Alhosseini et al.			✓	✓		0.6813	0.7318
SATAR			✓			0.8412	0.8642
BotRGCN	✓		✓	✓		0.8462	0.8707
GCN				✓		0.8670	/
GAT				✓		0.7750	/
HGT	✓			✓		0.8330	/
SimpleHGN	✓			✓		0.8690	/
RGT	✓		✓	✓		0.8664	0.8707
Follower View			✓	✓		0.8631	0.8785
Following View			✓	✓		0.8614	0.8768
Theme View	✓		✓	✓		0.8622	0.8767
Full Model	✓	✓	✓	✓	✓	**0.8749**	**0.8910**

The results demonstrate that:

- MV-GT consistently outperforms existing baseline models, including the newly proposed framework RGT [21]. We can achieve 0.8749 in accuracy and 0.8910 in F1-score.

- We propose for the first time to analyze the association between Twitter robots and users based on the underlying theme structure, which helps us explore the subtle differences between users and social robots under the theme structure.
- The proposed multi-view-based learning achieves the best performance on Twitter bot recognition, and these results demonstrate the necessity of fine-grained analysis of social networks through different views and the effectiveness of our method.

4.3 Model Architecture Study

In this paper, we propose a multi-view-based Twitter bot detection method. To demonstrate the validity of the model, multiple ablation experiments are set up to analyze the results.

Theme Structure Study

Figure 2 represents the perplexity curve drawn in order to find the optimal number of topics.

By maximizing the topic probability distribution, corresponding topics are assigned to each user tweet. Figure 3 shows the dimensionality reduction by t-SNE [27] to further visualize the topic distribution. The semantic associations between words in the same topic of tweets posted by users in social networks are large and small, due to the phenomenon of continuous distribution of nodes under the topic, which also illustrates that the topic structure in the proposed social network is one of the elements affecting user behavior.

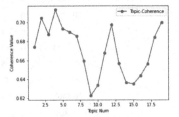

Fig. 2. Confusion curve

Fig. 3. Scatterplot of theme distribution

Multi-view Study

To further validate and analyze the effectiveness of multi-view learning, twitter bot detection is performed using a single view as well as a combination of multiple views by retaining the rest to separate only the multi-views (Fig. 4).

Learning from multiple views is superior to any single view learning. The results show that there is also an advantage of topic views over other views, which demonstrates the necessity of underlying topic structure for the analysis of user behavior. Further,

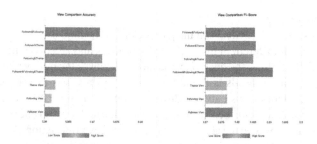

Fig. 4. Results of different view combinations

removing any of the views degrades the performance of the model, and multi-view learning can combine information from multiple views together so that the complex behavior of Twitter bots in social networks can be captured and understood more accurately.

View Fusion Study

An in-depth analysis of the multi-view fusion approach is performed to further demonstrate the effectiveness of the multi-view fusion method. Figure 5 shows the results of the comparison experiments.

Different approaches of multi-view fusion methods have advantages and disadvantages. The additive approach can effectively utilize the information of different views, but may lead to ambiguity and instability of feature representation; the averaging approach can fuse the information of different views, but will ignore the information of some views; the attention mechanism can focus on the important view information and improve the data representation, but may ignore the important view features and lead to the decrease of result accuracy. Therefore, on the basis of attention mechanism, adding connection operation and view interaction function can better retain the original feature information and utilize the interconnection between views, thus improving the performance of multi-view fusion. After comparative experimental analysis, the attention mechanism with connection operation and view interaction function is better than several other view fusion methods.

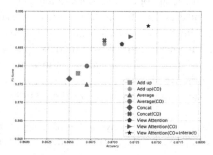

Fig. 5. Comparison experiment of view fusion methods CO means with connection operation Interact means with view interaction function

Parameter Sensitivity

We performed sensitivity analysis on the parameters of the model and plotted heat maps to show the effect of different number of attention heads and layers on the performance of the model. By analyzing the heat map, the model achieves the best performance when the number of attention heads is 8 and the number of layers is 8. It indicates that in social networks, relationships between different nodes may have different importance, and using multiple attention heads allows the model to focus on these relationships simultaneously and improve the accuracy of extracting key information. Also, increasing the number of layers of the model allows the model to understand and reason about the input data in a deeper way (Fig. 6).

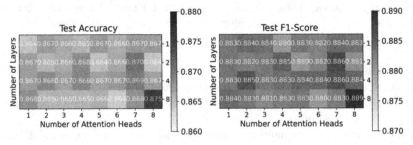

Fig. 6. Number of layers and number of attention heads

5   Conclusion

Social bot detection is an important and challenging task, which requires effective methods to automatically identify real users to cope with the changing social media users and related needs. In this paper, we propose a multi-view-based social bot detection framework, which analyzes the significant differences between users and social bots under social networks from a multi-view perspective. Based on this, we introduce a new topic structure through social cognitive theory to analyze the user behavior under this structure. Further, we also explore the multi-view fusion algorithm and improve it. We conduct extensive experiments on publicly available benchmarks to demonstrate the effectiveness of our model against state-of-the-art models. In addition, we explored and analyzed the effectiveness of topic structure and multiple views for social bot detection through ablation experiments. We plan to introduce more laws and theories of real-world user behavior in the future to deeply analyze the differentiation of users and social bots.

Acknowledgement. This article is supported by the National Key Research and Development Program of China (No. 2020YFC1523300), Innovation Platform Construction Project of Qinghai Province (2022-ZJ-T02).

References

1. Gorodnichenko, Y., Pham, T., Talavera, O.: Social media, sentiment and public opinions: evidence from #Brexit and #USElection. Eur. Econ. Rev. **136**, 1–22 (2021)
2. Grinberg, N., Joseph, K., Friedland, L., Swire-Thompson, B.: Fake news on Twitter during the 2016 U.S. presidential election. Science **363**(6425), 374–378 (2019)
3. Cresci, S.: A decade of social bot detection. Commun. ACM **63**(10), 72–83 (2020)
4. Cresci, S., Pietro, R.D., Petrocchi, M., Spognardi, A.: Fame for sale: efficient detection of fake twitter followers. Decis. Support. Syst. **80**, 56–71 (2015)
5. Lee, K., Eoff, B., Caverlee, J.: Seven months with the devils: a long-term study of content polluters on Twitter. In: 5th International AAAI Conference on Weblogs and Social Media, pp. 185–192. AAAI Press, Spain (2011)
6. Gupta, A., Lamba, H., Kumaraguru, P.: Faking sandy: characterizing and identifying fake images on Twitter during hurricane sandy. In: Proceedings of the 22nd international conference on World Wide Web companion. pp. 729–736. Association for Computing Machinery, Janeiro (2013)
7. Cai, C., Li, L., Zeng, D.: Detecting social bots by jointly modeling deep behavior and content information. In: Proceedings of the 2017 ACM on Conference on Information and Knowledge Management, pp. 1995–1998. Association for Computing Machinery, Singapore (2017)
8. Alhosseini, S.A., Tareaf, R.B., Najafi, P.: Detect me if you can: spam bot detection using inductive representation learning. In: Companion Proceedings of the 2019 World Wide Web Conference, pp. 148–153. Association for Computing Machinery, San Francisco (2019)
9. Zhao, C.S., Xin, Y., Li, X., Zhu, H., Yang, Y.: An attention-based graph neural network for spam bot detection in social networks. Appl. Sci. **10**(22), 1–15 (2020)
10. Yang, Y.G., Yang, R.Y., Cui, K.: RoSGAS: adaptive social bot detection with reinforced self-supervised GNN architecture search. ACM Trans. Web 1–32 (2022)
11. Chu, Z., Gianvecchio, S., Wang, H.: Detecting automation of Twitter accounts: are you a human, bot, or cyborg. IEEE Trans. Dependable Secure Comput. **9**(6), 811–824 (2012)
12. Bogdanova, D., Rosso, P., Solorio, T.: Exploring high-level features for detecting cyberpedophilia. Comput. Speech Lang. **28**(1), 108–120 (2014)
13. Heidari, M., Jones, J.H., Uzuner, O.: Deep contextualized word embedding for text-based online user profiling to detect social bots on Twitter. In: 2020 International Conference on Data Mining Workshops, Sorrento, Italy, pp. 480–487 (2020)
14. Wu, Y.H., Fang, Y.Z.: A novel framework for detecting social bots with deep neural networks and active learning. Knowl.-Based Syst. **211**(1), 1–18 (2020)
15. Mazza, M., Cresci, S., Avvenuti, M., Quattrociocchi, W.: RTbust: exploiting temporal patterns for botnet detection on Twitter. In: Proceedings of the 10th ACM Conference on Web Science, pp. 183–192. Association for Computing Machinery, Amsterdam (2019)
16. Ping, H., Qin, S.: A social bots detection model based on deep learning algorithm. In: 2018 IEEE 18th International Conference on Communication Technology, pp. 1435–1439. Chongqing (2018)
17. Lingam, G., Rout, R.R., Somayajulu, D.V.L.N.: Adaptive deep Q-learning model for detecting social bots and influential users in online social networks. Appl. Intell. **49**(11), 3947–3964 (2019)
18. Mesnards, N., Hunter, D.S., Hjouji, Z.E., Zaman, T.: Detecting bots and assessing their impact in social networks. Oper. Res. **70**(1), 1–22 (2021)
19. Wei, F., Nguyen, U.T.: Twitter bot detection using bidirectional long short-term memory neural networks and word embeddings. In: 2019 First IEEE International Conference on Trust, Los Angeles, pp. 101–109 (2019)

20. Feng, S., Wan, H., Wang, N., Luo, M.: BotRGCN: Twitter bot detection with relational graph convolutional networks. In: Proceedings of the 2021 IEEE/ACM International Conference on Advances in Social Networks Analysis and Mining, pp. 236–239. Association for Computing Machinery, Netherlands (2021)
21. Feng, S., Tan, Z., Li, R., Luo, M.: Heterogeneity-aware Twitter bot detection with relational graph transformers. In: Proceedings of the AAAI Conference on Artificial Intelligence, pp. 3977–3985. AAAI Press, California (2022)
22. Peng, H., Zhang, Y., Sun, H., Bai, X.: Domain-aware federated social bot detection with multi- relational graph neural networks. In: 2022 International Joint Conference on Neural Networks, Padua, Italy, pp. 1–8 (2022)
23. Li, Z., Liu, Z.L.: MV-GCN: multi-view graph convolutional networks for link prediction. IEEE Access 7, 176317–176328 (2019)
24. Xie, Y., Zhang, Y.Q., Gong, M.G.: MGAT: multi-view graph attention networks. Neural Netw. 132, 180–189 (2020)
25. Shi, Y., Huang, Z., Feng, S., Zhong, H.: Masked label prediction: unified message passing model for semi-supervised classification. In: International Joint Conferences on Artificial Intelligence Organization, pp. 1–7 (2020)
26. Feng, S., Wan, H., Wang, N., Li, J.: TwiBot-20: a comprehensive Twitter bot detection benchmark. In: Proceedings of the 30th ACM International Conference on Information & Knowledge Management, pp. 4485–4494. Association for Computing Machinery, Queensland (2021)
27. Van der Maaten, L., Hinton, G.: Visualising data using t-SNE[J]. J. Mach. Lear. Res. 9(11) (2008)
28. Yang, K. C., Varol, O., Hui, P. M.: Scalable and generalizable social bot detection through data selection. In: Proceedings of the AAAI Conference on Artificial Intelligence, pp. 1096–1103. AAAI Press, California (2020)
29. Feng, S., Wan, H., Wang, N., Li, J.: SATAR: a self-supervised approach to twitter account representation learning and its application in bot detection. In: Proceedings of the 30th ACM International Conference on Information & Knowledge Management, pp. 3808–3817. Association for Computing Machinery, Queensland (2021)
30. Kudugunta, S., Ferrara, E.: Deep neural networks for bot detection. Inf. Sci. 467, 312–322 (2018)

Scheduling Containerized Workflow in Multi-cluster Kubernetes

Danyang Liu[1], Yuanqing Xia[1(✉)], Chenggang Shan[2], Guan Wang[1],
and Yongkang Wang[1]

[1] School of Automation, Beijing Institute of Technology, Beijing 100081, China
{xia_yuanqing,wang_yk}@bit.edu.cn
[2] School of Artificial Intelligence, Zaozhuang University, Zaozhuang 277160, China

Abstract. Docker and Kubernetes have revolutionized the cloud-native technology ecosystem by offering robust solutions for containerization and orchestration workflows. This combination provides unprecedented speed, scalability, and efficiency in deploying and managing applications in distributed environments. However, when scheduling complex workflows across multi-cluster Kubernetes environments, existing workflow scheduling systems often fail to provide the necessary support. Integrating workflow scheduling algorithms with multi-cluster scheduling algorithms poses a complex and challenging problem. In this paper, we present a comprehensive framework known as the Containerized Workflow Engine (CWE), specifically designed for multi-cluster Kubernetes deployments. The CWE framework employs a two-level scheduling scheme, which combines the benefits of workflow containerization and establishes seamless connections between multi-cluster scheduling algorithms and multi-cluster Kubernetes environments. By integrating workflow scheduling algorithms with Kubernetes schedulers across Kubernetes environments, the CWE framework enables efficient utilization of resources and improved overall workflow performance. Compared to the state-of-the-art Argo workflows, CWE performs better in average task pod execution time and resource utilization.

Keywords: Workflow · Scheduling · Containerized

1 Introduction

Cloud infrastructure is continually evolving due to advancements in cloud-native technologies, hardware capabilities, networking enhancements, and the adoption of industry standards [21]. Cloud-native technologies, including containers, microservices, DevOps, Kubernetes [9], and other transformative practices such as serverless computing, infrastructure as code, and CI/CD, have revolutionized IT operations, maintenance, and development. Docker [5] and Kubernetes have emerged as prominent tools for cloud resource management, playing a significant role in the cloud-native technology ecosystem [13]. However, alternative solutions

© The Author(s), under exclusive license to Springer Nature Singapore Pte Ltd. 2023
E. Chen et al. (Eds.): BigData 2023, CCIS 2005, pp. 149–163, 2023.
https://doi.org/10.1007/978-981-99-8979-9_12

and platforms are available, such as Podma [3] and Apache Mesos [10], catering to specific use cases and requirements.

Kubernetes is an open-source container orchestration and management platform for automating containerized applications' deployment, scaling, and management. It provides a highly scalable, reliable, and user-friendly way to build, deploy, and manage applications across multiple hosts. Kubernetes defines the desired state of an application and automates tasks such as container creation, replication, restarts, and scaling, ensuring that the application runs consistently with the defined state. It also offers powerful service discovery and load balancing capabilities, supports horizontal auto-scaling, and enables developers and operations teams to efficiently build and manage modern distributed applications.

Single-cluster workflow scheduling [12] may face the risk of a single point of failure. For example, when the cluster experiences a failure, the workflow may be interrupted or halted. Furthermore, single-cluster workflow scheduling has limited resource utilization and cannot effectively handle load fluctuations. Multi-cluster workflow scheduling [22] offers better elasticity, scalability, and high availability than single-cluster workflow scheduling. Multi-cluster workflow scheduling can automatically adjust the cluster size based on workflow demands, ensuring elastic allocation and scalability of resources. Additionally, multi-cluster workflow scheduling can achieve resource isolation, meet geographical distribution and data locality requirements, and enhance the performance and efficiency of workflows.

Argo workflows (Argo) [2] is a powerful open-source workflow orchestration engine that focuses on managing and orchestrating containerized workloads in Kubernetes environments. The engine provides rich workflow orchestration capabilities, a visual interface, and close integration with Kubernetes. However, the current Argo scheduler has a significant drawback: it cannot schedule workflows across multi-cluster Kubernetes. This means that when it comes to orchestrating and scheduling workflows across multi-cluster, users need to implement cross-cluster scheduling logic themselves. This problem poses a challenging issue for containerized workflow [16,20] scheduling on Kubernetes and urgently requires an efficient framework closely integrated with Kubernetes to address this problem.

In this paper, we design a Containerized Workflow Engine, referred to as CWE, based on the further development of CWB [19]. The main objective of this system is to support workflow scheduling across multi-cluster Kubernetes and improve the execution efficiency of workflows by implementing a two-level scheduling scheme and containerized execution on Kubernetes. The CWE system consists of two components: Containerized Workflow Controller (CWC) and Containerized Workflow Scheduler (CWS). The CWC component can be deployed on high-performance hosts, and its primary function is to receive workflows and send them to the CWS components in multi-cluster. CWC implements a dual-channel mechanism, where the fast channel is used for quick workflow forwarding, while the slow channel ensures accurate routing of workflows to the schedulers. Additionally, CWC implements load balancing across clusters and within clusters. Specifically, when CWC receives an actual task, it decides which cluster to send the new task to based on the current workload of each cluster. CWC communicates with CWS to understand the load situation of CWS within the cluster, enabling it to decide whether to send the task to an existing CWS or start a new one. This design effectively addresses the pressure of large-scale cloud workflow tasks on the system. It avoids the predicament of a single CWS facing a sharp decline in performance or even failure to function properly due to massive requests. The CWS component is deployed in each Kubernetes cluster. To ensure smooth execution of workflow scheduling, CWS internally employs advanced workflow scheduling algorithms and utilizes the informer component to monitor Kubernetes resources. It also uses the Clint-go package to implement task container creation functionality. CWS utilizes the Goroutine mechanism to create concurrent task containers after the current task is completed for cases with multiple parallel successor tasks in a workflow. Furthermore, CWS handles data dependencies between task containers using the dynamic volume-sharing feature of StorageClass. Experimental results demonstrate that our CWE system exhibits better performance in terms of workflow execution efficiency. Compared to state-of-the-art technologies, CWE achieves a 31.61% improvement in enhancing workflow execution efficiency. Our contributions are summarized as follows:

- Design a framework for effectively managing containerized workflows within a Kubernetes environment. This framework incorporates a two-level scheduling scheme, allowing workflow management across multi-cluster Kubernetes.
- Implement a workflow injection module, CWC, and CWS. The workflow injection module is designed to handle the task injection process into the CWC during experiments. The primary role of the CWC is to transmit workflow tasks to the CWS within the Kubernetes cluster, taking into account the resource status of the cluster. The CWS is responsible for efficiently scheduling workflows within the Kubernetes environment.
- Provides a case study of containerized workflow in simulated production practice and presents a detailed performance analysis of CWE compared to other workflow scheduling solutions.

We have open-sourced the CWE. The source code is publicly available on GitHub at [4]

2   Related Work

As the standard container orchestration tool in the cloud-native era, Kubernetes provides rich and comprehensive support for developing the container application ecosystem. Its emergence offers powerful functionalities for the development and deployment of cloud-native applications and drives the continuous evolution of workflow engines. Airflow [1] is a platform to programmatically author, schedule, and monitor workflows. Airflow provides a user-friendly interface for defining, scheduling, and monitoring workflows as directed acyclic graph (DAG), offering features like task dependencies, error handling, and extensibility. Nextflow [7] is a bioinformatics workflow manager that enables the development of portable and reproducible workflows. Nextflow simplifies the creation and execution of scalable scientific workflows, supporting large-scale data and computational workloads with its DSL and containerization capabilities. Argo is an open-source container-native workflow engine hosted by Cloud Native Computing Foundation (CNCF). Argo enables the deployment and management of containerized applications in Kubernetes clusters, allowing users to define complex workflows as code with features such as templating and event-driven execution. Volcano [11], born in Huawei Cloud Native, is CNCF's first batch computing project. Volcano optimizes scheduling and resource management for batch and AI workloads on Kubernetes clusters, improving resource utilization and job performance through intelligent resource allocation and prioritization. These platforms empower organizations to streamline and automate their data processing and workflow management tasks, enhancing productivity and scalability.

Containerized workflow scheduling remains a relatively emerging research field. Existing technologies and tools, such as Airflow, Nextflow, Argo, and Volcano, primarily focus on workflow scheduling within a single Kubernetes cluster. However, their support for multi-cluster Kubernetes is not yet comprehensive enough. Consequently, a framework is needed to operate efficiently in a multicluster Kubernetes environment. Furthermore, Airflow and Nextflow were not originally intended as native workflow systems for Kubernetes, while Volcano was primarily focused on batch tasks. Presently, Argo is a cloud-native workflow engine specifically designed for Kubernetes. Therefore, in this paper, the experimental evaluation will primarily compare the submission method using Argo.

3   Design

This section provides a detailed explanation of the scientific workflow definition and the two-level scheduling scheme. We present the architectural design of the CWE and subsequently introduce the CWC and the CWS.

3.1   Scientific Workflow

In large-scale data processing tasks, the workflow [14,15] is typically described using a DAG to represent a distributed system application comprehensively. The relationships between tasks can be likened to edges in a DAG graph [23]. In addition to shared files, dependencies between tasks may involve data transfer, message queues, API calls, and other means. Container technologies [18] such as Docker provide a lightweight virtualization solution to encapsulate the execution environment and required resources for workflow tasks. The advantages of container technology include isolation, portability, and repeatability, utilizing container images as static snapshots of containers. In Kubernetes, a Pod is the smallest scheduling unit, serving as a logical deployable entity consisting one or more related containers, providing a shared network and storage environment.

The Kubernetes scheduler determines suitable nodes within the cluster to schedule Pods based on resource requirements, affinity rules, and scheduling policies.

3.2 Two-Level Scheduling Scheme

CWE and Kubernetes combine to implement a two-level scheduling scheme, as shown in Fig. 1. CWE serves as the interface that connects the workflow injector module and Kubernetes. Through the CWC module, the workflow scheduling algorithm is used to make scheduling decisions for workflows and distribute them to the CWS modules of different Kubernetes clusters. The CWS module is responsible for executing workflow tasks containerized to fully utilize cluster resources and improve the execution speed of workflows. The CWS module uses workflow scheduling algorithms to manage cluster resources efficiently, ensuring tasks are scheduled and executed based on task dependencies and resource requirements.

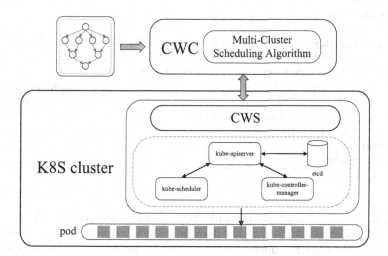

Fig. 1. Two-level scheduling scheme. The two-level scheduling scheme refers to Multi-cluster scheduling of CWC and workflow scheduling of CWS.

3.3 CWC Architecture

As shown in Fig. 2, CWC includes distributor module, pre-selector module, pressure evaluator module and state tracker module. Algorithm 1 shows the details. The function INITIALIZECLUSTER is responsible for the initialization of a cluster, involving the allocation of available resources and the computation of its initial score. To begin, the available resources of the cluster are gathered (line 1). Subsequently, the initial score of the cluster is calculated (lines 2 and 3). This initial score, along with the available resources, is then appended to the registry table (line 4) for future reference. In the subsequent function, TASKSCHEDUL-ING, the process of task allocation is orchestrated by systematically examining each cluster entry in the registry table. For each cluster under consideration, the algorithm first assesses whether the task's requirements can be accommodated by the cluster's available resources (lines 10 and 11). Following this, a predictive score for the cluster is computed (lines 11 and 12), which aids in the determination of its suitability for the given task. Ultimately, the task is assigned to the cluster that best matches with its needs.

Algorithm 1. Scheduling Algorithm

1: **function** INITIALIZECLUSTER(Cluster n, Available resources c_n, m_n, b_n, Total resources C_n, M_n, B_n)
2:    **if** Cluster n is new **then**
3:       $P_n = \alpha \frac{c_n}{C_n} + \beta \frac{m_n}{M_n} + \gamma \frac{b_n}{B_n}$ ▷Calculate initial cluster score
4:       Add cluster score and available resources to the registry table R
5:    **end if**
6: **end function**
7:
8: **function** TASKSCHEDULING(Registry table R, Task t, Task requirements c_t, m_t, b_t)
9:    **for** each cluster n in R **do**
10:       **if** Task requirements can be met by cluster n **then**
11:          $S_t = \alpha \frac{c_t}{C_n} + \beta \frac{m_t}{M_n} + \gamma \frac{b_t}{B_n}$ ▷Calculate task score
12:          $P_n = P_n - S_t$ ▷Update cluster score
13:          Assign task t to the cluster n
14:       **end if**
15:    **end for**
16: **end function**

---

a) Distributor Module: Responsible for sending workflows to CWS in multi-cluster Kubernetes. Through the distributor module, workflow tasks are intelligently allocated to different Kubernetes, resulting in optimized resource utilization and streamlined task execution. The module implements flexible resource allocation and load-balancing strategies, continuously adapting to demand variations. These capabilities enable higher concurrency, enhance system scalability, and improve overall performance.

b) Pre-Selector Module: Responsible for establishing a pre-selection table in advance, utilizing the current Kubernetes resource data obtained from CWS. This table provides precise information to the controller allocator, ensuring facilitating quick routing turnover and accurate workflow routing to the optimal scheduler. After scheduling workflows by the allocator module, the pre-selection table is updated in real-time according to the resource evaluation algorithm to correct each cluster's scoring and CWS load values. By predictive real-time update mechanism enables higher accuracy in pre-selection tables, improving the efficiency and reliability of workflow scheduling for CWC, resulting in improved performance.

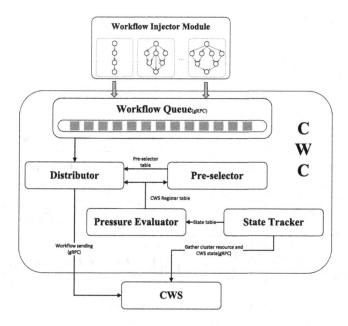

Fig. 2. Architecture of the CWC.

c) Pressure Evaluator Module: Responsible for continuously monitoring and evaluating the workload pressure of each CWS and Kubernetes cluster within the system and creating a corresponding workload registry table. Analyzing real-time and historical data calculates workload pressure scores, which indicate the level of resource utilization for each Kubernetes cluster. This information is used to optimize workload balancing and routing of workflows to ensure efficient task execution, maximize resource utilization, and enhance system performance. The module collaborates with the Pre-Selector and Distributor modules to analyze workload pressure data and make informed decisions regarding workflow distribution.

d) State Tracker module: Responsible for real-time monitoring and management of workflow and CWS statuses. It tracks the progress of workflows, ensuring their successful execution and handling failed workflows by rescheduling them. Furthermore, the module continuously monitors CWS to detect potential issues and updates the state table accordingly. Through active tracking of workflow execution, efficient management of failures, and maintenance of an accurate state table, this module significantly enhances the reliability and effectiveness of the CWS.

3.4 CWS Architecture

As shown in Fig. 3, CWS includes the scheduler, resource allocator, and tracker modules.

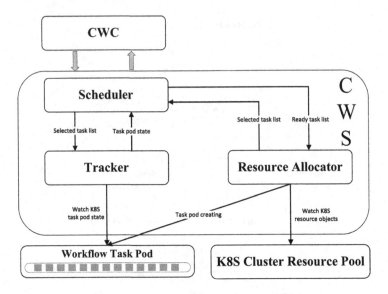

Fig. 3. Architecture of the CWS.

a) Scheduler module: Responsible for implementing critical algorithms for cloud workflow scheduling. The main objective was to efficiently allocate and manage workflow tasks submitted by users, ensuring optimal resource allocation and meeting personalized requirements. The Scheduler module analyzed task dependencies and resource demands, allocating them effectively among available Kubernetes cluster resources to achieve optimal execution efficiency and resource utilization. This module considered factors like task priority, data transmission between tasks, and resource utilization to formulate appropriate scheduling strategies.

b) Resource allocator module: Responsible for implementing the resource allocation functionality for workflow tasks. Its main functions include containerizing workflow tasks, monitoring Kubernetes resources using the informer component, creating task containers with the Clint-go package, creating concurrent task containers using the Goroutine mechanism after the current task is completed, handling data dependencies between task containers through dynamic volume-sharing using StorageClass, caching resource information locally to reduce API access pressure, and generating namespaces for achieving isolated environments for workflow resources. By effectively allocating and managing workflow tasks, optimizing resource utilization, meeting personalized requirements, and enhancing execution efficiency, this module ultimately improves the overall workflow performance.

c) Task Tracker module: Responsible for monitoring the execution status of cloud workflow task containers and providing real-time feedback to the scheduler to support the orderly execution of task containers. It detects the health of containers, collects and stores container log information, records the execution time of task containers, provides task progress updates, cleans up containers, and releases resources promptly after completing tasks.

3.5   Workflow Injection Module

The workflow injection module is an independent auxiliary module that operates separately from the CWE. Its primary functions include generating workflows, handling input requests from subsequent workflows, and transmitting workflow information to the CWC via gRPC. This module establishes the overall structure of workflow tasks and utilizes the Json method to inject configuration files containing task dependencies into the respective containers.

4   Experimental Evaluation

This section will evaluate the proposed CWE using various evaluation metrics and discuss its benefits compared to Argo.

4.1   Experimental Setup

To assess the performance of CWE, we have designed the workflow injection module. This module is containerized for deployment with CWC and CWE. Effective communication between these modules is facilitated through the gRPC mechanism.

The Kubernetes cluster used in our experiments consists of one master node and five worker nodes. Each node equips with a 6-core AMD EPYC 7742 2.2 GHz CPU and 8 GB of RAM, running Ubuntu 20.04 and Kubernetes v1.19.6 and Docker version 18.09.6 and Argo v3.2.9. CWC and workflow injector module are deployed on a high-performance virtual machine, and CWS is containerized and deployed into the Kubernetes cluster through Service and Deployment. In order to evaluate the performance of CWE across multi-cluster Kubernetes, we utilized a total of nine Kubernetes clusters. Due to Argo's lack of support for multi-cluster scheduling, we established a separate Kubernetes cluster comprising three master nodes and forty-five worker nodes for Argo.

4.2   Workflow Example

In order to validate the application scalability of CWE, we have tailored a customized workflow that encompasses all the node-dependent characteristics of the DAG diagram, accommodating more intricate scenarios. The workflow task program employs resource loads to simulate workflow tasks in real-world production practice.

a) Workflow Topology: We utilize a DAG diagram to represent the workflow, constructing an experimental example encompassing all the typical characteristics of such a diagram. As shown in Fig. 4, this workflow comprises seven tasks featuring branching, crossover, and merging elements. Based on the interdependencies among task nodes, the scheduling algorithm employed for this workflow follows a top-down approach, ensuring tasks are scheduled topologically.

*b) Workflow Containerization:*Taking inspiration from [17], we adopt a Python application as a workflow task and utilize the Stress tool [6] to emulate CPU and memory usage within a defined timeframe. To facilitate this, we employ the Docker engine to package the Python application into a task Image file. This task Image file is subsequently stored in either a local Harbor [8] or a remote Docker Hub repository, and its image address is initialized within the workflow injection module. Furthermore, container parameters can be imported into the task container, specifying CPU cycles, memory allocation, and duration, all of which contribute to determining the runtime of the task pod. The task involves CPU forking and memory allocation operations, executed over 15 s. Within the JSON file, we specify the task pod's resource requests and resource limit parameters as 1000 milli cores for CPU and 512Mi for memory. It is worth noting that the requests and limits fields share the same parameter values.

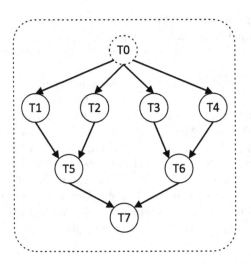

Fig. 4. Workflow topology diagram.

4.3 Results and Analysis

In order to evaluate the effectiveness of CWE, our first step is to verify the workflow execution efficiency on multi-cluster Kubernetes using CWE. Subsequently, we will compare CWE and Argo, focusing on workflow execution efficiency and CPU usage rate. We will now describe the two methods for workflow submission.

- CWE: We employ the containerized method to deploy CWE. CWC and work-flow injector module are deployed on a high-performance virtual machine, while CWS is deployed within each Kubernetes cluster. We use a JSON file describing a DAG to represent the workflow task dependency relationship. After going through the workflow injector module, compress the JSON file using Snappy and submit it to CWC via gRPC.
- Argo: We deployed the Argo Workflow image in the Kubernetes cluster using the official YAML file provided by Argo. Similar to the CWE, we employ the same JSON file and leverage a workflow injection module to convert it into a YAML format that Argo can recognize and then submit to Argo.

a) Workflow task execution efficiency: We package Docker images for CWC, CWS, and workflow injection modules. We define YAML files for RBAC, StorageClass, and CWS. CWC and workflow injection modules are containerized and deployed on high-performance virtual machines with a Docker engine. The YAML files are deployed in the Kubernetes cluster, where CWS are scattered and scheduled to the cluster nodes as pods. The components communicate with each other using gRPC.

As is shown in Fig. 5, The execution time for each group of workflows has been averaged across ten experiments. The execution time of the CWE workflow task was determined by subtracting the start time of the workflow injector from the *successedWorkflows* metric. Similarly, the execution time of the Argo workflow task was determined by subtracting the start time of the workflow injector from the *Successfully* metric found in the log of the Argo workflow-controller pod. The CWE takes 133.3 s to receive 100 workflow tasks from the workflows injection module until the workflows pod is execution completed, 350.1 s to execution completed 500 workflow tasks, and 825.8 s to execution completed 1000 workflows tasks. The Argo takes 121.2 s to receive 100 workflow tasks from the workflows injection module until the workflows pod is execution completed, 431.1 s to execution completed 500 workflow tasks and, 1086.89 s to execution completed 1000 workflows tasks. During the initial execution of 100 workflows, CWE and Argo exhibited similar execution times, indicating sufficient resources within the Kubernetes cluster. However, as the workload increased to 500 workflows, CWE experienced a 23.13% decrease in execution time compared to Argo. This disparity can be attributed to inadequate resources within the Kubernetes cluster. When executing 1000 workflows, Argo generated workflow task pods exclusively within its namespace, resulting in a significant accumulation of pods. This accumulation ultimately led to the restart of the Argo workflow-controller pod. Consequently, CWE experienced a 31.61% decrease in execution time compared to Argo.

In addition, upon comparing the execution of 500 workflows with that of 1000 workflows, it becomes evident that CWE outperforms in handling larger-scale workflows. It validates that the CWE is a framework for a large-scale workflow scheduling tool for multi-cluster Kubernetes.

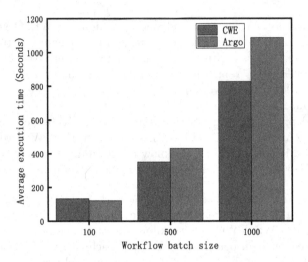

Fig. 5. Average execution time of workflow.

b) Resource Usage Comparison: This section aims to utilize Prometheus to capture the state changes of underlying resources in a Kubernetes cluster under different numbers of workflows to showcase the CPU utilization characteristics of two engines, CWE and Argo. To ensure accurate performance comparisons for CPU usage, it is crucial to address the substantial impact caused by frequent resource fluctuations. In order to mitigate this influence, we meticulously configure our experimental environment to eliminate any additional workloads that could affect performance measurements. To enhance the stability and resource allocation efficiency of the Kubernetes cluster, the Master node is intentionally excluded from participating in pod scheduling and workload, focusing solely on its core administrative tasks.

Figure 6, Fig. 7 show the CPU usage rate of the CWE and Argo over the lifecycle of 100 and 500 workflows. When executing 100 workflows in two Kubernetes clusters with the same number of nodes, the CPU utilization curves of the two workflow engines are similar. However, the CPU utilization curve shows significant fluctuations when executing 500 workflows using the Argo engine. This could be due to the fact that when the Argo workflow engine executes a large number of workflows simultaneously, all the workflow pods are launched in the same namespace, resulting in a drastic drop in system performance and the inability to schedule workflows properly. Regardless of the type of Kubernetes cluster mode, the available number of CPU cores is 270000 milli. Under no load conditions, the CPU utilization of the Kubernetes cluster components is 0.7%. After injecting workflows, the CWS is launched and requires 2 CPU cores. Therefore, after completing workflow execution, the CPU utilization of CWE will remain at 7%. Due to the prolonged time required by Argo to clean up completed pods, which marks the end of workflow execution, there has been a significant performance degradation issue

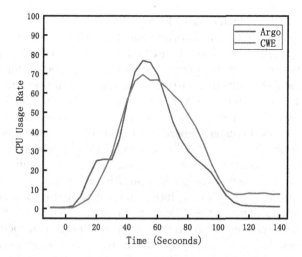

Fig. 6. CPU Usage Rate for 100 Workflows.

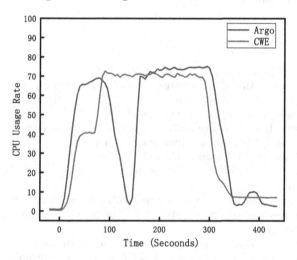

Fig. 7. CPU Usage Rate for 500 Workflows.

when a large number of workflow injections occur. We designed CWE that assigns a separate namespace for each workflow task, enabling resource isolation and more efficient cleanup of completed pods. In a multi-cluster Kubernetes environment, our CWE has better performance.

5   Conclusion

In this paper, our CWE has successfully achieved efficient workflow task scheduling across multi-cluster Kubernetes. CWE offers comprehensive workflow management functionalities, including task definition, dependency management, and execution sequencing. It also employs intelligent distributed scheduling strategies to allocate tasks to different clusters based on resource availability and workload conditions, thus enhancing the overall system efficiency and performance.

Our experimental results demonstrate a significant improvement in the workflow scheduling throughput of CWE compared to the state-of-the-art single-cluster workflow scheduling engine, Argo, with an approximate increase of around 31.61% in multi-cluster Kubernetes scenarios. This indicates the superior scheduling performance and scalability of CWE in multi-cluster environments.

In conclusion, as a multi-cluster workflow scheduling engine, CWE holds promising prospects for a wide range of applications. By providing flexible workflow management functionalities and intelligent distributed scheduling strategies, CWE significantly improves the efficiency and performance of workflow task scheduling in multi-cluster environments. Future research can focus on further refining the scheduling algorithms, optimizing resource management strategies, and expanding the capabilities of CWE to cater to the growing demands of containerized workflow tasks.

References

1. Apache airflow (2023). https://airflow.apache.org/
2. Argo-workflows - github (2023). https://github.com/argoproj/argo-workflows
3. The best free and open source container tools (2023). https://podman.io/
4. CWE - github (2023). https://github.com/liudy093/CWE
5. Develop faster. Run anywhere (2023). https://www.docker.com/
6. Linux man page (2023). https://linux.die.net/man/1/stress
7. Nextflow (2023). https://www.nextflow.io/
8. Our mission is to be the trusted cloud native repository for kubernetes (2023). https://goharbor.io/
9. Production-grade container orchestration (2023). https://kubernetes.io/
10. Program against your datacenter like it's a single pool of resources (2023). https://mesos.apache.org/
11. Volcano - github (2023). https://github.com/volcano-sh/volcano
12. Adhikari, M., Amgoth, T., Srirama, S.N.: A survey on scheduling strategies for workflows in cloud environment and emerging trends. ACM Comput. Surv. (CSUR) **52**(4), 1–36 (2019)
13. Bernstein, D.: Containers and cloud: from LXC to docker to Kubernetes. IEEE Cloud Comput. **1**(3), 81–84 (2014)
14. Bharathi, S., Chervenak, A., Deelman, E., Mehta, G., Su, M.H., Vahi, K.: Characterization of scientific workflows. In: 2008 Third Workshop on Workflows in Support of Large-Scale Science, pp. 1–10. IEEE (2008)
15. Deelman, E., Gannon, D., Shields, M., Taylor, I.: Workflows and e-science: an overview of workflow system features and capabilities. Futur. Gener. Comput. Syst. **25**(5), 528–540 (2009)
16. Hobson, T., Yildiz, O., Nicolae, B., Huang, J., Peterka, T.: Shared-memory communication for containerized workflows. In: 2021 IEEE/ACM 21st International Symposium on Cluster, Cloud and Internet Computing (CCGrid), pp. 123–132. IEEE (2021)

17. Klop, I.: Containerized workflow scheduling (2018)
18. Pahl, C.: Containerization and the PaaS cloud. IEEE Cloud Comput. **2**(3), 24–31 (2015)
19. Shan, C., Wang, G., Xia, Y., Zhan, Y., Zhang, J.: Containerized workflow builder for Kubernetes. In: 2021 IEEE 23rd International Conference on High Performance Computing & Communications; 7th International Conference on Data Science & Systems; 19th International Conference on Smart City; 7th International Conference on Dependability in Sensor, Cloud & Big Data Systems & Application (HPCC/DSS/SmartCity/DependSys), pp. 685–692. IEEE (2021)
20. Shan, C., Xia, Y., Zhan, Y., Zhang, J.: KubeAdaptor: a docking framework for workflow containerization on Kubernetes. Futur. Gener. Comput. Syst. **148**, 584–599 (2023)
21. Varghese, B., Buyya, R.: Next generation cloud computing: new trends and research directions. Futur. Gener. Comput. Syst. **79**, 849–861 (2018)
22. Wang, Y.R., Huang, K.C., Wang, F.J.: Scheduling online mixed-parallel workflows of rigid tasks in heterogeneous multi-cluster environments. Futur. Gener. Comput. Syst. **60**, 35–47 (2016)
23. Zheng, C., Tovar, B., Thain, D.: Deploying high throughput scientific workflows on container schedulers with makeflow and mesos. In: 2017 17th IEEE/ACM International Symposium on Cluster, Cloud and Grid Computing (CCGRID), pp. 130–139. IEEE (2017)

A Study of Electricity Theft Detection Method Based on Anomaly Transformer

Shufen Chen[1], Yikun Yang[1], Shuaiying You[1], Wenbin Chen[2],
and Zhigang Li[1(✉)]

[1] Shihezi University, Shihezi 832003, China
lizhigang1998@163.com
[2] Xinjiang Tianfu Information Co., Ltd, Shihezi 832003, China

Abstract. Electricity theft not only disrupts normal electricity consumption but also poses a significant security threat to the power system. The widespread deployment of smart meters has led to the collection of massive amounts of electricity consumption data, which can help identify electricity theft. However, the challenge of detecting electricity theft is heightened by the category imbalance in the electricity consumption data collected. In this study, we address this problem by using ADASYN resampling technology to balance data categories, and then develop a model based on Anomaly Transformer (AT) to identify electricity theft by analyzing historical data that deviates from normal patterns following a theft. The model uses an attention mechanism to calculate and extract the series-association between power consumption data streams, and a Gaussian kernel to calculate the priori-association of the relative temporal distance between power consumption data points and their neighbors. We validate the proposed model using the SGCC dataset, and our experimental results demonstrate high accuracy, precision, F1-score, and AUC values.

Keywords: Electricity theft detection · Anomaly Transformer · Deep Learning · Non-technical losses · Smart meters

1 Introduction

Line losses in power systems can be divided into technical losses (TL) and non-technical losses (NTL) [7]. The technical losses caused by aging lines and unreasonable network structure have been well solved with the transformation of the power grid, so the non-technical losses become more prominent. Among them, power theft is one of the essential causes of non-technical losses.

Advanced metering infrastructure (AMI) is a technology that connects customers and electric utilities through a wide area communication network [21,31]. Smart meters collect customers' electricity consumption data at a high frequency to provide a reference for subsequent work by electric utilities. However, attackers can use specialized equipment to hack smart meters, tamper with electricity

E. Chen et al. (Eds.): BigData 2023, CCIS 2005, pp. 164–180, 2023.
https://doi.org/10.1007/978-981-99-8979-9_13

consumption records, or attack meters [11]. This may result in a failure to extract reference information from the collected electricity consumption data. In China, incomplete statistics from the national power department indicate that electricity theft causes an annual loss of more than 10 billion. This highlights the significant economic losses and safety hazards associated with electricity theft, including the risk of major accidents such as fires that threaten the safety and stability of the electric power system.

At present, outdated detection methods, unlabeled collected electricity consumption data, and exponential growth in the amount of data make it difficult to detect electricity theft in a timely manner and extract valuable information from the data. However, the emergence of artificial intelligence offers a potential solution to this problem. Researchers can use AI to identify electricity theft by detecting patterns in the electricity consumption data following a theft, which differs from the normal consumption patterns.

The detection of anomalies in electricity use is a primary research focus in the analysis of electricity usage. At present, conventional Machine Learning (ML) and Deep Learning (DL) techniques are widely used to identify instances of electricity theft.

Many researchers have utilized traditional ML methods for detecting power theft, as these methods are straightforward to comprehend and quick to train [2,12,17,28]. Traditional ML methods can be classified as supervised learning and unsupervised learning, depending on whether the model training process requires labeled data or not. Random Forest (RF) and K-Nearest Neighbor (KNN) are classical supervised learning algorithms. RF consists of multiple decision trees, which have strong resistance to noise, but the presence of multiple similar decision trees can obscure the actual classification results. The literature [6,13] proposes RF algorithms for time series anomaly detection. The former algorithm defines anomalies based on the complexity of the decision tree and utilizes probability to randomly cut the tree across feature dimensions, resulting in better anomaly detection performance than RF. The latter algorithm employs a combination of Convolutional Neural Networks (CNN) and RF for detecting power theft. KNN can perform both classification and regression tasks, but its computational cost increases significantly as the number of features in the dataset grows too large. In literature [3], a KNN method for detecting electricity theft was designed by analyzing historical electricity consumption data and performing feature extraction, resulting in a high accuracy rate. However, supervised learning anomaly detection methods are limited in this field due to the challenges posed by handling data labels and selecting appropriate parameters.

In ML, electricity theft detection using unsupervised learning methods is typically categorized into clustering-based, tree-based, and other methods. In literature [16], k-means with Local Outlier Factor (LOF) are used to select and calculate the degree of anomalies for outliers, respectively. In literature [10], an improved DBSCAN algorithm is used to detect anomalies in time series data with seasonality, which is more effective than the basic DBSCAN algorithm in identifying anomalies. Although many scholars have achieved anomaly detection through clustering, the selection of radius in the clustering process is an important issue that cannot be ignored. Decision Tree (DT) is a classical tree-based method for anomaly detection, and many researchers [18,19,23] have developed DT-based models for detecting electrical theft. However, most of these models ignore the correlation between electricity consumption data. Unsupervised learning algorithms cannot assess the quality of the resulting model parameters because they do not use abnormal user electricity usage data labels during the training process, and traditional machine learning methods rarely take into account temporal information, which limits their applicability in electricity anomaly detection scenarios [27]. However, traditional ML relies heavily on feature engineering, which involves manually designing feature extraction schemes and incurs high pre-processing costs. In addition, traditional ML requires high-quality data and has a weak generalization ability, which makes it challenging to adapt to the electricity theft detection scenario. Deep learning-based methods for power theft detection, on the other hand, can automatically extract features and are highly adaptable to large datasets, thereby compensating for the limitations of traditional ML. In Deep Learning, Neural Networks serve as an end-to-end architecture that automatically learns the relationship between input and output. In literature [4,9], electrical anomaly detection models based on Recurrent Neural Networks (RNN) have been developed. The former proposed a method that combines an Encoder-Decoder framework with RNN. The latter utilized RNN for electricity consumption prediction and compared the prediction results with actual consumption, using the difference between the two and a preset threshold to detect anomalies in power data. Several researchers [8,15,26] have proposed CNN-LSTM models for identifying customer electricity theft. These models leverage the Convolutional Neural Network's (CNN) ability to automatically extract features and the Long Short-Term Memory (LSTM) network's superior performance on continuous time-series datasets. Literature [20,24] proposes electricity consumption anomaly detection models based on LSTM, the latter focusing more on the former with seasonality and monthly trends. With continuous research, the Transformer model has been proposed, and its important component, the self-attention mechanism, has been widely applied in various fields due to its excellent capabilities. In literature [5], a power theft detection model was constructed based on a multi-headed attention mechanism, and a binary input channel was introduced to identify missing values in the power consumption dataset. Experimental results show that this model achieves an AUC of 0.92 on the SGCC dataset. In literature [25,29], anomaly detection models were constructed using the attention mechanism combined with Convolutional Neural Networks (CNN). The literature [22] proposes a multivariate time series anomaly detection model (OmniAnomaly) based on the combination of VAE (Variational AutoEncoders) and GRU, which is robust to various devices. The literature [30] proposes a CAE-M method, which uses a deep convolutional autoencoder as the feature extraction module, an attention-based bidirectional long- and short-term memory model and an autoregressive model as the prediction module, and calculates the error value of the objective function for each test data, and uses the size of the error value as the basis for judging the normal and abnormal data categories. Experimental results show that the introduction of the attention mechanism improves the detection capability of the model. Despite scholarly research demonstrating the feasibility of DL in power theft detection scenarios, the following problems still exist:

1) Data imbalance: Electricity theft data is usually scarce, which means that the number of negative samples in the training data may be very small. In this case, the model may be overfitted, resulting in a high false alarm rate.
2) The complexity of the model structure: DL models need to be trained several times or stacked with multiple layers of neural networks if they want high performance, which will take a lot of time and resources.

In this paper, the Anomaly Transformer(AT)-based electricity theft detection model is used to identify customers' electricity theft behavior in response to the problems of the above studies. Firstly, the electricity consumption data are processed using resampling techniques to make the proportion of positive and negative categories in the electricity consumption data suitable for use in the AT model; secondly, the model uses a multi-headed attention mechanism to calculate and extract the serial association between electricity consumption data streams [14] and uses a Gaussian kernel to calculate the priori-association of the relative temporal distance between electricity consumption data points and neighboring points [27]. Finally, the difference between the priori-association and the serial-association is used to discriminate the electricity theft data.

The overall structure of this paper is as follows: Section 2 provides a detailed analysis of the customer power consumption model and an introduction to the data pre-processing methods used. Section 3 presents the AT model in detail. Section 4 gives the experimental results. Finally, Sect. 5 gives a summary of the paper.

2 Characteristic Analysis and Data Expansion

2.1 Data Analysis

This study utilized the daily consumption records of 42,372 customers from January 1, 2014, to October 30, 2016, which were publicly available from the State Grid Corporation (SGCC) [1]. The details are presented in Table 1, which highlights the significant imbalance between normal customers and electricity theft customers. Given that data are collected on a daily basis, it is possible for power consumption data to contain missing or abnormal values due to malfunctioning smart meters, unreliable transmission of measurement data, unplanned maintenance of the system, and storage issues [32]. To address this issue, missing data are filled using the linear interpolation method, except when the values are , in which case they are filled with 0.

$$f(x_i) = \begin{cases} \frac{x_{i-1}+x_{i+1}}{2} & x_i \in \text{NaN}, \\ x_{i-1}, x_{i+1} \notin \text{NaN} \\ \\ 0 & x_i \in \text{NaN}, \\ x_{i-1} \text{or} x_{i+1} \in \text{NaN} \\ \\ x_i & x_i \notin \text{NaN}, \end{cases} \tag{1}$$

x_i denotes the amount of electricity consumption data for a certain customer on a certain day, or Nan if x_i is a non-numeric type. The outliers in the electricity consumption data are recovered using the 3δ principle, and the 3δ principle is:

$$f(x_i) = \min(x_i, \text{mean}(\mathbf{X}) + 3\text{std}(\mathbf{X})) \tag{2}$$

In this paper, $\mathbf{X}$ represents a day-by-day vector of electricity consumption data, where x_i denotes the electricity consumption data of day i in $\mathbf{X}$. The mean and standard deviation of $\mathbf{X}$ are denoted by "mean" and "std", respectively. Only positive deviations are considered, as the electricity consumption data of each customer is always greater than or equal to 0. To facilitate faster convergence of the neural network and avoid numerical problems, normalizing the electricity consumption data is essential. In this study, we normalize the data using the linear normalization method:

$$f(x_i) = \frac{x_i - \min(\mathbf{X})}{\max(\mathbf{X}) - \min(\mathbf{X})} \tag{3}$$

where min(X), max(X) are the minimum and maximum values in X, respectively.

Table 1. Distribution of SGCC dataset.

Description	Value
Data collecting time slot	1 January 2014–31 October 2016
Type of data	Times series
All customers	42372
Normal customers	38757
Electricity theft customers	3615
Electricity theft ratio/Normal	0.09

2.2 Data Expansion Mechanism

When there is a significant imbalance between electricity theft data and normal electricity consumption data, the model is highly sensitive to normal electricity consumption data, resulting in unsatisfactory classification results. To address this issue, this paper employs the resampling technique ADASYN to augment the customer electricity consumption data, which involves the following steps:

Step1: To determine the Euclidean distance between each minority class sample x and all other minority class samples, perform the following calculation:

Step2: The density of each minority class sample x is determined by calculating the number of other minority class samples that are within a certain distance from it.

Step3: To determine the number of synthetic samples to be generated for each minority class sample x, the distance ratio is calculated by computing the ratio of its distance to all other minority class samples to the number of majority class samples whose distance falls within a certain range.

Step4: The number of synthetic samples to be generated for each minority class sample is determined by calculating the distance ratio.

Step5: To generate synthetic samples for each minority class sample x, a sample y is randomly selected from its k-nearest neighbors, and an interpolation is performed between them to create a new synthetic sample. This process is repeated until the desired number of synthetic samples is obtained.

2.3   Feature Analysis

This study utilizes the SGCC dataset to analyze the characteristics of normal electricity consumption patterns and electricity theft patterns. Figure 1 illustrates the 30-day electricity consumption of normal users and electricity theft users. It can be observed that the electricity consumption of electricity theft users has no sharp peaks and is consistently low compared to normal users. Analyzing the users' electricity consumption pattern by week reveals an obvious periodicity for normal users, as shown in Fig. 2(a). Conversely, there is no clear periodicity in the weekly electricity consumption of electricity theft users, and the temporal characteristics disappear, as shown in Fig. 2(b). Hence, the electricity consumption data are segmented strictly according to a 7-day interval and converted to the following data format:

$$\mathbf{X}_i = \begin{bmatrix} x_{1,1} & \cdots & x_{1,7} \\ \vdots & \ddots & \vdots \\ x_{n,1} & \cdots & x_{n,7} \end{bmatrix} \tag{4}$$

X_i denotes the electricity consumption data of the ith user, and $x_{n,1}$ denotes the electricity consumption of the user on Monday of the nth week.

3   Electricity Theft Detection Model

3.1   Electricity Theft Detection Methods

The electricity theft detection model based on Anomaly Transformer (AT) [27] consists of an anomaly detection module with stacked L-layers and a feedforward neural network, as shown in Fig. 3.

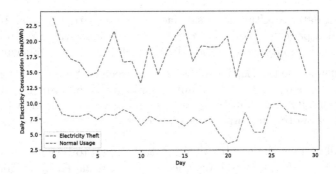

Fig. 1. 30-day electricity consumption line graph of normal users and electricity theft users.

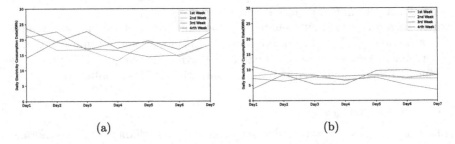

(a)   (b)

Fig. 2. (a) is a line graph of electricity consumption data for normal users and (b) is a line graph of electricity consumption data for theft users.

The anomaly detection of power consumption data consists of two branches: the priori-association and series-association of power consumption data. The difference between these two associations is defined as the association difference, which is used as a criterion for subsequent anomaly detection. The input and output relationship between the anomaly detection module of the alternately stacked l-layer and the feedforward neural network can be expressed as follows:

$$\mathcal{Z}^l = \text{Layer-Norm} \left(\text{Anomaly-Attention} \left(\mathcal{X}^{l-1} \right) + \mathcal{X}^{l-1} \right) \tag{5}$$

$$\mathcal{X}^l = \text{Layer-Norm} \left(\text{Feed-Forward} \left(\mathcal{Z}^l \right) + \mathcal{Z}^l \right) \tag{6}$$

where l denotes the current layer and $x \in \mathbb{R}^{N \times d_{model}}$ (N is the input timing length and d is the timing dimension). z^l denotes the hidden representation of the l layer.

Priori-Association: The priori-association incorporates a learnable Gaussian kernel function, with the center being the index of the corresponding time point. This module utilizes the single peak of the Gaussian distribution and focuses more on neighboring points in the electricity data.

$$P^l = \text{Rescale}\left(\left[\frac{1}{\sqrt{2\pi}\sigma_i}\exp(-\frac{|j-i|^2}{2\sigma_i^2})\right]_{i,j\in\{1,\cdots,N\}}\right) \tag{7}$$

where $\sigma \in \mathbb{R}^{N\times 1}$ is the parameter of the Gaussian kernel function, and P^l is the a priori association of the l layer, computed by the Gaussian kernel to represent the association weight of the i time point with the j point.

Series-Association: The series-association is derived from a multi-headed attention calculation in the standard Transformer. The sequential association of a point is represented by the distribution of attention weights for that point in the corresponding row of the attention matrix. This module is designed to extract associations from the original sequence and enable the model to adaptively capture the most effective associations.

$$\mathbf{Q}, \mathbf{K}, \mathbf{V}, \sigma = X^{l-1}\mathbf{W}_Q^l, X^{l-1}\mathbf{W}_K^l, X^{l-1}\mathbf{W}_V^l, X^{l-1}\mathbf{W}_\sigma^l \tag{8}$$

$$S^l = \text{Softmax}\left(\frac{\mathbf{Q}\mathbf{K}^T}{\sqrt{d_{\text{model}}}}\right) \tag{9}$$

$$Y^l = S^l V \tag{10}$$

where $\mathbf{Q}, \mathbf{K}, \mathbf{V} \in \mathbb{R}^{N\times d_{model}}, \sigma \in \mathbb{R}^{N\times 1}$ are query, key, value and Gaussian kernel function, W is the matrix of parameters to be trained, respectively, S^l is the sequence association, and $Y^l \in \mathbb{R}^{N\times d_{model}}$ is the hidden representation of the l layer Anomaly Transformer. In the multi-headed attention mechanism, if there are h heads, $\sigma \in \mathbb{R}^{N\times h}$ and $Y^l \in \mathbb{R}^{N\times d_{model}}$ is connected by $\{Y^l \in \mathbb{R}^{N\times d_{model}}\}_{1\leq n\leq h}$.

Association Discrepancy: The priori-association of anomalies and the series-association that focuses more on neighboring points exhibit a slight difference. Conversely, the attention map of normal sequences has a global focus characteristic and is distributed over non-neighboring points, resulting in a substantial difference between these two associations compared to the distinct priori-association of single peaks. Based on this characterization of association differences as a metric of anomalies, the degree of difference between a priori associations and sequential associations is expressed using the symmetrized KL scatter formula and averaged over association differences from multiple layers, which can be expressed as:

$$\text{AssDis}(P, S; X) = \left[\frac{1}{L}\sum_{l=1}^{L}\left(\text{KL}(P_{i;}^l \| S_{i,}^l) + \text{KL}(S_{i,}^l \| P_{i,}^l)\right)\right]_{i=1,\cdots,N} \tag{11}$$

where $l \in \{1, \cdots, L\}$ and $\text{KL}(\cdot \| \cdot)$ denotes the calculation of the discrete distribution between P^l, S^l for each row of data, $AssDis(P, S; X)$ represents the

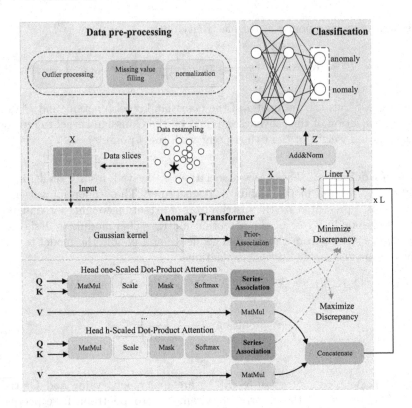

Fig. 3. General architecture of the model.

point-by-point difference between the prior association of sequence X and the sequence association. Anomalous points have smaller differences than normal points.

Minimax Strategy: When $\lambda > 0$, minimizing the loss function results in maximizing the association difference. However, directly maximizing the association difference can cause the σ of the Gaussian kernel to become excessively small, leading to a meaningless prior distribution. Therefore, the problem is solved by employing a very large minima strategy while changing the direction of λ in the Loss to realize the switching of the max-min learning direction, which is formulated as follows:

$$\mathcal{L}_{Total}\left(X, P, S, \lambda; X\right) = \|X - \widehat{X}\|_F^2 - \lambda \times \| \text{AssDis}(P, S; X)\|_1 \qquad (12)$$

$$\text{Minimize Phase:} \mathcal{L}_{Total}\left(\widehat{X}, P, S_{\text{detach}}, -\lambda; X\right) \qquad (13)$$

$$\text{Maximize Phase:} \mathcal{L}_{Total}\left(\widehat{X}, P_{\text{detach}}, S, \lambda; X\right) \qquad (14)$$

where $\mathcal{L}_{Total}$ is the loss function, $\widehat{X}$ is the reconstruction of the original electricity consumption data sequence, and λ is the balance term.

In the Minimize Phase, the priori-association P is optimized to approximate the serial association S. This process enables the priori-association to adapt to the electricity consumption data pattern while avoiding the σ parameter of the Gaussian kernel from becoming too small.

In the Maximize Phase, the difference between associations is maximized by optimizing the serial association S. This process focuses the serial association more on non-adjacent points, making the reconstruction of anomalies more challenging.

Anomaly Score: The smaller AssDis and the larger reconstruction loss result in a higher Anomaly score, as follows:

$$\text{AnomalyScore}(X) = \text{Softmax}\left(-\text{AssDis}\left(P, S; X\right)\right) \odot \left[\left\|X_{i,:} - \widehat{X}_{i,:}\right\|_2^2\right]_{i=1,\cdots,N} \tag{15}$$

3.2   Electricity Theft Detection Specific Process

The theft electricity detection method studied in this article completes the process of theft electricity detection in three steps:

Step1: Data pre-processing involved two steps. First, missing values were filled, and outliers were removed from the acquired customer electricity consumption data. Second, the electricity consumption data were sliced by week and normalized.

Step2: Data enhancement was performed by resampling the pre-processed data using the ADASYN resampling method. This method increased the number of positive and negative class samples, resulting in a relatively balanced dataset.

Step3: The Gaussian kernel function and Multi-Head Attention are used to extract the priori-association and series-association of electricity consumption data, respectively. The difference between them is increased using the great minima strategy, which enables anomaly detection through error reconstruction.

4   Experimental Evaluation

4.1   Data Expansion Performance Evaluation

The ADASYN resampled electricity consumption data is visualized and compared with the original electricity consumption data. Since a customer's electricity consumption data consists of 1029 features, the T-SNE algorithm is used for visualization and analysis. Figure 4 shows the distribution of the original electricity consumption data on the left and the distribution of the ADASYN resampled electricity consumption data on the right. The comparison reveals that the electricity theft data generated by ADASYN after resampling is very similar to the distribution of the real data. Thus, the electricity theft data generated by ADASYN is deemed reliable.

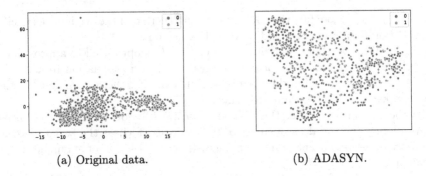

(a) Original data.   (b) ADASYN.

Fig. 4. Comparison of the distribution of raw power consumption data with the power consumption data after resampling by ADASYN.

4.2   Dataset Preparation

This study strictly analyzes the electricity consumption characteristics of customers on a weekly basis. Therefore, the SGCC dataset from January 6, 2014 (Monday) to October 30, 2016 (Sunday) will be used, and the 1029 days of electricity consumption data for each customer will be divided into 147 weeks. The 42,375 customers will be divided into training and test sets in an 8:2 ratio.

4.3   Evaluation Metrics

In this paper, Accuracy, Precision, Recall, F1-Score, and AUC evaluation metrics are used to examine the performance of Anomaly Transformer-based power theft models, calculated as follows:

$$Accuracy = \frac{TP + TN}{TP + FP + TN + FN} \tag{16}$$

$$Precision = \frac{TP}{TP + FP} \tag{17}$$

$$Recall = \frac{TP}{TP + FN} \tag{18}$$

$$F1 - score = \frac{2 * Precision * Recall}{Precision + Recall} \tag{19}$$

In binary classification, TP represents the number of normal users detected as electricity thieves, while FP represents the number of normal users misidentified as electricity thieves. TN represents the number of normal users correctly identified as normal users, and FN represents the number of electricity thieves misidentified as normal users.

Testing the electricity theft model using only the above evaluation metric is not sufficient. Based on the experience of other studies, the AUC evaluation metric is the best performance metric for detecting binary classification tasks.

The AUC is the area under the ROC curve, which represents the probability that a pair of samples (one positive and one negative) is randomly selected and then classified correctly by the trained classifier. The formula for AUC is as follows:

$$AUC = \frac{\sum_{i \in positiveclass} R_i - \frac{M(M+1)}{2}}{M * N} \tag{20}$$

where, R_i denotes the serial number of the ith data (the probability is sorted from largest to smallest), M, N denotes the number of normal electricity users and electricity theft users respectively.

4.4   Model Parameters

This study compares the proposed electricity theft detection model with both deep learning methods (CNN, LSTM, CNN-LSTM, OmniAnomaly [22] , CAE_M [30]) and traditional machine learning methods (KNN, DT) using the specific parameters listed in Table 2. The experiments were conducted in the same environment with the same dataset to test the classification effectiveness of the models.

Table 2. Model Parameters.

Method	Parameters
DT	max_depth = 5
KNN	n_neighbors = 3
CNN	kernel_size = 3, filters = 64, batch_size = 147
CNN-LSTM	kernel_size = 3, filters = 64, units = 200, batch_size = 147
OmniAnomaly	kernel_size = 3, window_size = 5, batch_size = 147
CAE_M	kernel_size = 3, window_size = 5,batch_size = 147
AT	kernel_size = 3, filters = 64, n_heads = 8, L = 3, d_model = 512, batch_size = 147

4.5   Analysis of Results

Tables 3 and 4 show the experimental results of electricity theft detection obtained by AT as well as DT, KNN, CNN, CNN-LSTM, OnmiAnomaly and CAE_M in two resampling methods, SMOTE and ADASYN.

Table 3 and 4 illustrate that the detection performance of models trained on the electricity consumption dataset is significantly improved after SMOTE and ADASYN resampling compared to the original dataset. The AUC values for most of the electricity theft models using the original data are below 0.6, with the AUC value for DT only reaching 50.8%. However, after resampling the

Table 3. Comparison of raw electricity consumption data and SMOTE resampled electricity consumption data.

Method	Original					SMOTE				
	Accuracy	Precision	Recall	F1-Score	AUC	Accurary	Precision	Recall	F1-Score	AUC
DT	0.844	0.100	0.104	0.102	0.510	0.854	0.832	0.888	0.859	0.847
KNN	0.902	0.321	0.141	0.196	0.534	0.739	0.662	0.993	0.795	0.739
CNN	0.909	0.431	0.237	0.306	0.603	0.945	0.913	0.988	0.949	0.878
CNN-LSTM	0.911	0.250	0.065	0.100	0.651	0.959	0.931	0.992	0.961	0.922
OmniAnomaly	0.980	0.901	0.862	0.881	0.926	0.980	0.903	0.862	0.882	0.927
CAE_M	0.974	0.891	0.793	0.839	0.892	0.978	0.901	0.839	0.869	0.915
Proposed	0.985	0.901	0.931	0.914	0.938	0.988	0.900	0.965	0.932	0.978

Table 4. Comparison of raw electricity consumption data and power consumption data after resampling by ADASYN.

Method	Original					ADASYN				
	Accurary	Precision	Recall	F1-Score	AUC	Accurary	Precision	Recall	F1-Score	AUC
DT	0.844	0.100	0.104	0.102	0.510	0.843	0.826	0.869	0.847	0.837
KNN	0.902	0.321	0.141	0.196	0.534	0.760	0.682	**0.996**	0.809	0.747
CNN	0.909	0.431	0.237	0.306	0.603	0.954	0.927	0.992	0.958	0.929
CNN-LSTM	0.911	0.250	0.065	0.100	0.651	0.966	0.950	0.983	0.966	0.961
OmniAnomaly	0.980	0.901	0.862	0.881	0.926	0.978	0.860	0.885	0.873	0.936
CAE_M	0.974	0.891	0.793	0.839	0.892	0.959	0.821	0.666	0.736	0.826
Proposed	0.985	0.901	0.931	0.914	0.938	**0.984**	**0.990**	0.979	**0.984**	**0.982**

original data using the SMOTE and ADASYN methods, most of the electricity theft models have AUC values higher than 80%.

This section presents a comparison of experimental results for the raw power consumption data, power consumption data after SMOTE resampling, and power consumption data after ADASYN resampling:

1) The electricity consumption data was processed using the SMOTE resampling technique and compared to the original data. As shown in Table 3, the AUC values of DT, KNN, CNN, CNN-LSTM, OmniAnomaly, CAE_M and AT improved by 0.337, 0.205, 0.275, 0.271, 0.001, 0.023 and 0.040, respectively;

2) The electricity consumption data was processed using the ADASYN resampling technique and compared to the original data. As shown in Table 4, the AUC values of DT, KNN, CNN, CNN-LSTM, OmniAnomaly, and AT increased by 0.327, 0.213, 0.326, 0.310, 0.010 and 0.044, respectively. However, the CAE_M model using the ADASYN resampled dataset is worse than the original data, and the reasons are still being further analyzed;

3) The electricity consumption data was subjected to both SMOTE resampling and ADASYN resampling. The experimental results indicate that the ADASYN resampling technique yields higher AUC values for KNN, CNN, CNN-LSTM, OmniAnomaly and AT compared to the SMOTE resampling technique by 0.008, 0.051, 0.039, 0.009 and 0.004, respectively.

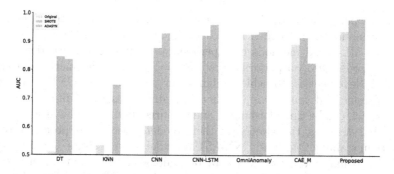

Fig. 5. Comparison of AUC obtained experimentally from raw data, after resampling the data.

As shown in Fig. 5 and Fig. 6, compared to other comparative models, the theft model used in this paper achieves optimal values of Accuracy, Precision, F1-score, and AUC under the ADASYN resampling method. Therefore, the results suggest that the ADASYN method is more suitable for the theft model used in this paper.

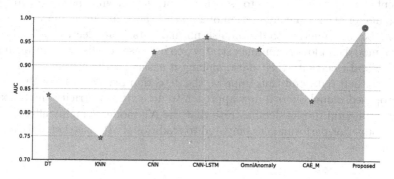

Fig. 6. Comparison of AUC obtained experimentally after resampling data by ADASYN.

The Anomaly Transformer (AT)-based power theft model offers three key advantages. Firstly, multidimensional feature extraction is no longer limited to manual methods or dependent on more complex network structures. Secondly, feature extraction relies on the dot product of matrices, enabling accelerated computing using hardware such as GPUs. Finally, utilizing the difference between a priori and serial correlations to distinguish normal data from power theft data simplifies outlier differentiation.

The analysis of experiments reveals that traditional ML methods can rapidly process electricity consumption data and have straightforward model structures. However, traditional ML methods overlook the correlation between data attributes when processing electricity usage data, which can impact the detection performance of the model. Deep learning (DL)-based methods outperform traditional ML-based methods in experimental results. The method that uses an Autoencoder also outperforms the traditional ML-based method in terms of results. Nevertheless, to achieve high accuracy, CNN and CNN-LSTM may require deeper and more complex network structures or additional training rounds, which can be time-consuming and resource-intensive. This study demonstrates that training the electricity theft model three times produces better results than training CNN and CNN-LSTM models twenty times. OnmiAnomaly and CAE_M have more complexity, more model structure and number of trainings compared to AT. However, AT obtained higher AUC than OnmiAnomaly and CAE_M. To summarize, the Anomaly Transformer (AT) method demonstrates higher effectiveness in electricity theft detection compared to other models, based on the same dataset.

5   Conclusion

This study presents an Anomaly Transformer (AT) model that can effectively address the issue of electricity theft. As outliers are primarily related to neighboring points and it is challenging to establish a correlation with the entire sequence, the AT model utilizes a combination of a multi-headed attention mechanism and Gaussian kernel function to distinguish normal data from electricity theft data. The AT model employs a minimax strategy to increase the distance between normal data and electricity theft data, making it easier for the classifier to identify electricity theft data and thus improve the detection performance of the model. The proposed model offers a new approach to detecting electricity theft. Extensive experimental results demonstrate that the AT model is highly effective and exhibits superior performance compared to traditional ML based models and DL based models.

References

1. http://www.sgcc.com.cn/
2. Arif, A., Alghamdi, T.A., Khan, Z.A., Javaid, N.: Towards efficient energy utilization using big data analytics in smart cities for electricity theft detection. Big Data Res. **27**, 100285 (2022)
3. Aziz, S., Naqvi, S.Z.H., Khan, M.U., Aslam, T.: Electricity theft detection using empirical mode decomposition and k-nearest neighbors. In: 2020 International Conference on Emerging Trends in Smart Technologies (ICETST), pp. 1–5. IEEE (2020)
4. Fengming, Z., Shufang, L., Zhimin, G., Bo, W., Shiming, T., Mingming, P.: Anomaly detection in smart grid based on encoder-decoder framework with recurrent neural network. J. China Univ. Posts Telecommun. **24**(6), 67–73 (2017)
5. Finardi, P., et al.: Electricity theft detection with self-attention. arXiv preprint arXiv:2002.06219 (2020)

6. Guha, S., Mishra, N., Roy, G., Schrijvers, O.: Robust random cut forest based anomaly detection on streams. In: International Conference on Machine Learning, pp. 2712–2721. PMLR (2016)
7. Haq, E.U., Pei, C., Zhang, R., Jianjun, H., Ahmad, F.: Electricity-theft detection for smart grid security using smart meter data: a deep-CNN based approach. Energy Rep. **9**, 634–643 (2023)
8. Hasan, M.N., Toma, R.N., Nahid, A.A., Islam, M.M., Kim, J.M.: Electricity theft detection in smart grid systems: a CNN-LSTM based approach. Energies **12**(17), 3310 (2019)
9. Hollingsworth, K., et al.: Energy anomaly detection with forecasting and deep learning. In: 2018 IEEE International Conference on Big Data (Big Data), pp. 4921–4925. IEEE (2018)
10. Jain, P.K., Bajpai, M.S., Pamula, R.: A modified DBSCAN algorithm for anomaly detection in time-series data with seasonality. Int. Arab J. Inf. Technol. **19**(1), 23–28 (2022)
11. Jokar, P., Arianpoo, N., Leung, V.C.: Electricity theft detection in AMI using customers' consumption patterns. IEEE Trans. Smart Grid **7**(1), 216–226 (2015)
12. Kong, X., Zhao, X., Liu, C., Li, Q., Dong, D., Li, Y.: Electricity theft detection in low-voltage stations based on similarity measure and DT-KSVM. Int. J. Electr. Power Energy Syst. **125**, 106544 (2021)
13. Li, S., Han, Y., Yao, X., Yingchen, S., Wang, J., Zhao, Q.: Electricity theft detection in power grids with deep learning and random forests. J. Electr. Comput. Eng. **2019**, 1–12 (2019)
14. Li, Y., Zhang, L., Lv, Z., Wang, W.: Detecting anomalies in intelligent vehicle charging and station power supply systems with multi-head attention models. IEEE Trans. Intell. Transp. Syst. **22**(1), 555–564 (2020)
15. Madhure, R.U., Raman, R., Singh, S.K.: CNN-LSTM based electricity theft detector in advanced metering infrastructure. In: 2020 11th International Conference on Computing, Communication and Networking Technologies (ICCCNT), pp. 1–6. IEEE (2020)
16. Peng, Y., et al.: Electricity theft detection in AMI based on clustering and local outlier factor. IEEE Access **9**, 107250–107259 (2021)
17. Punmiya, R., Choe, S.: Energy theft detection using gradient boosting theft detector with feature engineering-based preprocessing. IEEE Trans. Smart Grid **10**(2), 2326–2329 (2019)
18. Saeed, M.S., Mustafa, M.W.B., Sheikh, U.U., Khidrani, A., Mohd, M.N.H.: Electricity theft detection in power utilities using bagged CHAID-based classification trees. J. Optim. Ind. Eng. **15**(2), 67–73 (2022)
19. Saeed, M.S., Mustafa, M.W., Sheikh, U.U., Khidrani, A., Mohd, M.N.H.: Theft detection in power utilities using ensemble of CHAID decision tree algorithm. Sci. Proc. Ser. **2**(2), 161–165 (2020)
20. da Silva, A., Guarany, I., Arruda, B., Gurjão, E.C., Freire, R.: A method for anomaly prediction in power consumption using long short-term memory and negative selection. In: 2019 IEEE International Symposium on Circuits and Systems (ISCAS), pp. 1–5. IEEE (2019)
21. Singh, N.K., Mahajan, V.: End-user privacy protection scheme from cyber intrusion in smart grid advanced metering infrastructure. Int. J. Crit. Infrastruct. Prot. **34**, 100410 (2021)
22. Su, Y., Zhao, Y., Niu, C., Liu, R., Sun, W., Pei, D.: Robust anomaly detection for multivariate time series through stochastic recurrent neural network. In: Proceed-

ings of the 25th ACM SIGKDD International Conference on Knowledge Discovery & Data Mining, pp. 2828–2837 (2019)

23. Tehrani, S.O., Moghaddam, M.H.Y., Asadi, M.: Decision tree based electricity theft detection in smart grid. In: 2020 4th International Conference on Smart City, Internet of Things and Applications (SCIOT), pp. 46–51. IEEE (2020)

24. Wang, X., Zhao, T., Liu, H., He, R.: Power consumption predicting and anomaly detection based on long short-term memory neural network. In: 2019 IEEE 4th International Conference on Cloud Computing and Big Data Analysis (ICC-CBDA), pp. 487–491. IEEE (2019)

25. Xia, R., Gao, Y., Zhu, Y., Gu, D., Wang, J.: An attention-based wide and deep CNN with dilated convolutions for detecting electricity theft considering imbalanced data. Electr. Power Syst. Res. **214**, 108886 (2023)

26. Xia, Z., Zhou, K., Tan, J., Zhou, H.: Bidirectional LSTM-based attention mechanism for CNN power theft detection. In: 2022 IEEE International Conference on Trust, Security and Privacy in Computing and Communications (TrustCom), pp. 323–330. IEEE (2022)

27. Xu, J., Wu, H., Wang, J., Long, M.: Anomaly transformer: time series anomaly detection with association discrepancy. arXiv preprint arXiv:2110.02642 (2021)

28. Yan, Z., Wen, H.: Electricity theft detection base on extreme gradient boosting in AMI. IEEE Trans. Instrum. Meas. **70**, 1–9 (2021)

29. Zhang, Y., Ji, Y., Xiao, D.: Deep attention-based neural network for electricity theft detection. In: 2020 IEEE 11th International Conference on Software Engineering and Service Science (ICSESS), pp. 154–157. IEEE (2020)

30. Zhang, Y., Chen, Y., Wang, J., Pan, Z.: Unsupervised deep anomaly detection for multi-sensor time-series signals. IEEE Trans. Knowl. Data Eng. **35**, 2118–2132 (2021)

31. Zheng, K., Chen, Q., Wang, Y., Kang, C., Xia, Q.: A novel combined data-driven approach for electricity theft detection. IEEE Trans. Industr. Inf. **15**(3), 1809–1819 (2018)

32. Zheng, Z., Yang, Y., Niu, X., Dai, H.N., Zhou, Y.: Wide and deep convolutional neural networks for electricity-theft detection to secure smart grids. IEEE Trans. Industr. Inf. **14**(4), 1606–1615 (2017)

Application and Research on a Large Model Training Method Based on Instruction Fine-Tuning in Domain-Specific Tasks

Yawei Zhang$^{(\boxtimes)}$, Changsheng Li, Xindong Wang, and Bin Zhang

China Unicom Software Research Institute, Data Middle Platform Research
and Development Business, Hong Kong, China
{zhangyw944,lics187,wangxd335,zhangb162}@chinaunicom.cn

Abstract. The potential of large-scale models to enhance industrial productivity and catalyze societal progress is undeniable. However, inherent challenges-such as lengthy training cycles and the demand for advanced computational resources-remain daunting. Given recent advancements in computational adaptability, this paper introduces a systematic approach to effectively fine-tune these models for domain-specific tasks. Our method encompasses three key phases: (1) a thorough analysis of domain-specific business needs and data acquisition; (2) precise task segmentation, designing standardized instruction formats to construct a fine-tuning dataset, and subsequently fine-tuning the large-scale models; (3) rigorous model validation using a test dataset. Through these steps, we effectively fine-tuned our training using 5,000 data instances and validated our results with an additional 1,000 test instances. To complement our study, we provide a comparative analysis of different training techniques and assess the fine-tuning results on four prominent open-source models. The conclusions drawn offer valuable insights for the future application of large-scale models in specialized domains and pave the way for further research and applications.

Keywords: Large-scale model · Instruction fine-tuning · Vertical task

1 Introduction

In recent years, large-scale pre-trained language models, commonly referred to as "large models," have attracted significant attention in the field of natural language processing (NLP). In 2018, OpenAI released the first iteration of the Generative Pretrained Transformer (GPT) model [1,8], marking the commencement of the "pre-training" era in NLP. However, despite the technology behind GPT is advanced, it did not immediately garner widespread attention. Instead, Google's Bidirectional Encoder Representations from Transformers (BERT) model [2,3] took center stage. Yet, OpenAI remained undeterred, continuing on the technical trajectory of GPT and releasing GPT-2 and GPT-3.

E. Chen et al. (Eds.): BigData 2023, CCIS 2005, pp. 181–194, 2023.
https://doi.org/10.1007/978-981-99-8979-9_14

Of particular note is the GPT-3 model, which boasts a massive 175 billion parameters and introduced the concept of "prompting" [4] for the first time. This innovative interaction mode offers three main advantages: firstly, it reduces the need for model fine-tuning, thereby lowering the costs of deploying the model for new tasks; secondly, it provides users the flexibility to interact with the model, enabling a wide range of applications and swift experiment iterations; lastly, it offers a simplified interaction mechanism for non-expert users. As a result, the introduction of prompts has greatly enhanced the practicality and versatility of GPT-3.

However, upon deep evaluation of GPT-3's performance, researchers found that large models still possess inherent issues typical of deep learning models, such as poor robustness, limited explainability, and constrained inferencing capabilities. It was only with the advent of ChatGPT [5–7] that perceptions about large models fundamentally shifted. These models harness vast amounts of data for knowledge extraction and learning [9], producing models with billions of parameters, heralding a new era in AI research.

Yet, ChatGPT, being a generative general-purpose large language model, cannot be directly applied to specific vertical domains, such as medicine, law, and finance. To adapt large models for specialized domain tasks, fine-tuning is imperative. This is because these vertical domains possess their unique terminologies and knowledge structures, which might not be fully covered in the training datasets of large models, and these models might struggle to adapt to the data distribution of these verticals. For instance, they may misunderstand professional terminologies or be unfamiliar with specific business content, leading to unintended errors in real-world applications. More crucially, ensuring the accuracy and reliability of the inferences drawn by these large models in critical tasks, such as medical diagnoses or addition, deletion, modification, and query operations in finance.

To address these challenges, this paper introduces an instruction fine-tuning method for training large models, aimed at enabling their application in vertical domains with minimal data. Our approach primarily revolves around Supervised Fine-Tuning (SFT), a method involving the pre-training of a neural network model on a source dataset and then creating a new neural network model by replicating the design and parameters of the source model (excluding the output layer). Fine-tuning can aid large models in better addressing domain-specific business needs, reduce potential biases and misconceptions, lower pre-training costs [4,10], and enhance performance on domain-specific tasks.

In this paper, leveraging instruction-based fine-tuning as a foundation and using operator data as a case study, we delve into the processes of data collection, standardizing instruction formats, and fine-tuning the parameters of large models to complete specific tasks in the operator domain, namely NL2SQL (Natural Language to SQL). NL2SQL is a task that involves converting natural language queries into their corresponding SQL (Structured Query Language) format, which is crucial for database querying and management. This task is pivotal in the operator domain, where rapid, accurate database queries are essential for efficient operations. By nature, it demands an intricate understanding of both natural language semantics and SQL syntax, a feat challenging for generic LLMs.

The efficacy of fine-tuning emerges prominently when addressing the complexities of the NL2SQL task. Through fine-tuning, the model acquires a nuanced understanding of domain-specific terminologies and contexts, which is instrumental in translating intricate natural language queries into their SQL counterparts. Moreover, the regimented syntax of SQL necessitates meticulous precision; any deviations can engender substantial discrepancies in database outputs. Fine-tuning ensures that the model meticulously adheres to SQL's stringent syntax, yielding queries that are not only syntactically accurate but also optimized for performance. Beyond mere syntactical considerations, the semantic integrity of a query remains paramount. The fine-tuning process imbues the model with the capability to discern subtle semantic distinctions in natural language queries, ensuring that the resultant SQL representations faithfully capture the intent of the original queries.

We fine-tuned the large model using 5,000 data and tested it with 1,000 data instances. The experimental results met our expectations, further validating that general-purpose large language models after fine-tuning, can effectively handle and address domain-specific challenges.

2   Related Work

On March 17th, 2023, OpenAI, Open Research, and the University of Pennsylvania published the latest research paper titled "GPTs are GPTs: An Early Look at the Labor Market Impact Potential of Large Language Models" [11]. . The authors believe that in the current development trend, large models are gradually becoming a universal technology, which will have profound potential impacts on the labor market. The paper explores the specific effects of LLMs (Large Language Models) on various professions and industries. However, large models have shown certain limitations in their application to vertical domains, such as misunderstanding of professional vocabulary and unfamiliarity with business content. These can lead to unexpected errors in practical applications. To address the errors produced by large models in vertical domain applications, we propose a method of instruction fine-tuning for training large models. This method can enable the application of large models in vertical domains based on a small amount of data. See Fig. 1

The primary training method for instruction fine-tuning is SFT (Supervised Fine-Tuning) [12–14]. SFT involves pre-training a neural network model on a source dataset, then creating a new neural network model, copying the design and parameters of the source model (except for the output layer). During fine-tuning, a new output layer is added to the target model and trained on the target dataset. SFT leverages the parameters and structures of the pre-trained models to speed up the training process and enhance the model's performance on vertical domain-specific tasks.

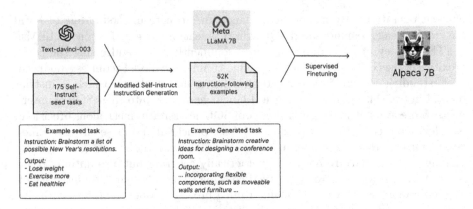

Fig. 1. Instruction Finetune Architecture

The leading methods in supervised fine-tuning currently include LoRA [15] (Low-Rank Adaptation of Large Language Models), P-tuning v2 [16], and Freeze [17]. They all possess unique advantages and characteristics.

2.1 LoRA

LoRA [15,17], which stands for Low-Rank Adaptation of Large Language Models, is a distinctive fine-tuning technique. Its fundamental principle involves freezing the pre-trained model's weight parameters and adding additional network layers to the model, training only these newly added layer parameters. A significant advantage of this method is that, due to the fewer added parameters, not only does the cost of fine-tuning decrease significantly, but similar effects to tuning the full model parameters can be obtained. LoRA freezes pre-trained model weights and injects the decomposition matrices of trainable rank into each layer of the Transformer architecture. For instance, for the pre-trained weight matrix W_0, its updates are constrained by using low rank decomposition to represent the latter. During model fine-tuning, W_0 is frozen, and only parameters A and B are fine-tuned. This substantially reduces the number of parameters to be fine-tuned compared to tuning all parameters.

Among the main advantages of LoRA are that pre-trained model parameters can be shared, used to build many small LoRA modules for different tasks. By freezing the shared model and replacing matrices A and B, tasks can be switched efficiently, significantly reducing storage requirements and the costs of switching between multiple tasks. When using an adaptive optimizer, LoRA improves fine-tuning performance and reduces the hardware threshold for fine-tuning by three times, as there is no need to calculate gradients and save too many model parameters. The low rank decomposition adopts a linear design approach, which enables the merging of trainable parameter matrices with frozen parameter matrices during deployment, without introducing inference delays compared to fully fine-tuning methods. LoRA does not conflict with various other fine-tuning methods and can be combined with other fine-tuning methods. The schematic diagram of LoRA is as follows: See Fig. 2:

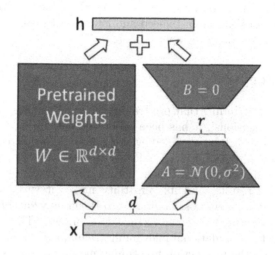

Fig. 2. LoRA Schema

2.2 P-Tuning

P-tuning v2 [16] is another fine-tuning method for large language models, an improved version of P-tuning v1, drawing inspiration from prefix-tuning. Compared to P-tuning v1, P-tuning v2 introduces tunable parameters at the beginning of every layer, while P-tuning v1 only fine-tuned the first layer. The improvements in P-tuning v2 include removing the Re-paramerization training acceleration method, adopting multi-task learning optimization, and abandoning the Verbalizer's vocabulary Mapping. Instead, it reutilizes [CLS] and character labels, enhancing its versatility, making it adaptable to sequence labeling tasks.

2.3 Freeze Fine-Tuning

Freeze [17] is another efficient fine-tuning technique for large language models. The core idea of this method is "parameter freezing." During the fine-tuning process, only a subset of the parameters is trained, while all others are frozen. Specifically, the Freeze method only fine-tunes the full connected layer parameters of the last few layers in the Transformer model. Experiments have shown that shallow layers of the Transformer mainly extract surface features, while deeper layers tend to extract semantic features. Thus, by only fine-tuning the full connected layers of the last few layers, the Freeze method manages to maximize the fine-tuning effects of large language models efficiently.

The emergence of these three fine-tuning methods not only enriches the technical means of fine-tuning large language models but also provides more choices for different task requirements and resource constraints. The flexibility and universality of LoRA make it suitable for various complex fine-tuning scenarios; the parameter efficiency and innovation of P-tuning v2 enable it to perform well on models of different scales; the fast speed and low cost of Freeze make it an ideal choice under resource-constrained situations.

3   Methodology

Due to previous work finding that pre-trained language models have a relatively lower 'intrinsic dimensions', it has been observed that during task adaptation, effective learning can still be obtained even if randomly projected into a smaller subspace. Because LoRA effectively adds a small parameter module to learn the changes, it reduces training costs, accelerates training speeds, efficiently optimizes iterative results, and its versatility makes it suitable for a variety of complex fine-tuning scenarios. Compared to P-tuning v2 and Freeze, LoRA is more efficient in adapting to specific vertical domain tasks. Therefore, this paper chooses LoRA as the foundational fine-tuning technique [18]. The large model is fine-tuned based on the instruction fine-tuning method of LoRA and applied to vertical domain tasks. applied to vertical domain tasks.

Suppose that the pre-train matrix is $W_0 \in \mathbb{R}^{d \times k}$, its updates can be represented by:

$$W_0 + \Delta W = W_0 + BA,$$

$$\text{where } B \in \mathbb{R}^{d \times k}, A \in \mathbb{R}^{r \times k}, \text{ and } r \ll min(d, k). \tag{1}$$

During initialization, matrix A is initialized using random Gaussian values, while matrix B is initialized with zeros. Throughout the training process, W_0 remains frozen and does not receive any gradient updates. In contrast, both A and B contain trainable parameters. Both W_0 and $\Delta W = BA$ multiply with the same input. Their respective output vectors are summed coordinate-wise. Furthermore, the output dimensions of W_0 and ΔW are consistent. As a result, after training is complete, merging the parameters simply involves element-wise addition at corresponding positions.

$$h = W_0 x + \Delta W x = W_0 x + BA x \tag{2}$$

During the inference process, one only needs to reintegrate the modifications back into the original model, ensuring that there is no additional latency.

$$W = W_0 + BA \tag{3}$$

In short, the LoRA training mode uses low-rank decomposition to simulate parameter changes, thereby indirectly training large models with a minimal amount of parameters. This approach enhances training speed and reduces the hardware threshold for training. See Fig. 3

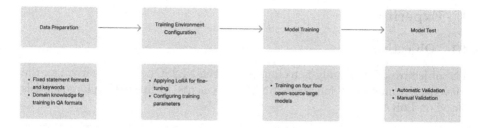

Fig. 3. Manual Verification process

In instruction fine-tuning, fields like 'instruction' are used to define downstream tasks, 'input' to specify user inputs, and 'output' to denote model outputs. Consequently, we establish a dataset format for instruction fine-tuning. Depending on the different 'instruction', different tasks can be defined. From this, it can be seen that we need to design template formats and sentence structures for instruction fine-tuning data based on specific tasks.

The primary objective of this paper is to implement the NL2SQL (Natural Language to SQL) capability through fine-tuning large models. This means using natural language descriptions to query data in a database and further supporting search functionalities based on question similarity, including search history and trending searches. The specific experimental goal of this paper is solely NL2SQL, converting natural language into executable SQL statements. The experimental data originates from the telecommunications domain and has undergone desensitization.

For domain-specific knowledge in telecommunications, such as proprietary terms like "billing cycle" and "indicators," general large language models might not comprehend effectively. To enable the big model to complete specific tasks in this field, we first enable the big model to understand proprietary terms. Based on this understanding, we use fixed statement formats and keywords as instruction bases, specifically for SQL table lookup, and develop specific instruction statements that include table and column names. This method not only ensures accurate and speedy database lookups but also reduces ambiguities often encountered by general models. By associating specific instruction statements with database structures, the approach offers consistent, reliable responses, enhancing user experience. Furthermore, it presents a scalable solution that can be extended to various sectors, providing businesses with data-driven insights and cost savings. The dataset comprises question texts and corresponding SQL statements. Detailed examples will be presented in the following Sect. 4.5.2

4   Experiment

4.1   Objective

The objective of this study is to implement the NL2SQL functionality, i.e., the conversion from natural language to SQL statements, by fine-tuning large models based on LoRA.

4.2   Dataset

For this experiment, we utilized a dataset provided by a telecommunications operator, comprising a total of 5,000 records. A specific example of the dataset is depicted in the following Table 1.

Table 1. Dataset: 5,000 data entries from a specific telecommunications provider

账期	月份	省份ID	省份名称	省份排序	地市ID	地市名称	地市排序	渠道类型
202210	202210	19	山西	25	V0140100	太原市	288	社会实体
202210	202210	19	山西	25	V0140100	太原市	288	自有实体
202210	202210	19	山西	25	V0140100	太原市	288	电子渠道
202210	202210	19	山西	25	V0140100	太原市	288	合计
202210	202210	19	山西	25	V0140100	太原市	288	政企渠道
202210	202210	19	山西	25	V0140200	大同市	289	政企渠道
202210	202210	19	山西	25	V0140200	大同市	289	合计
202210	202210	19	山西	25	V0140200	大同市	289	电子渠道

4.3   Fine-Tuning Pre-trained Models

To accomplish this task, we selected the following four pre-trained models for fine-tuning:

- Chinese-Llama-7B [19]
- ChatGLM2-6B [20]
- RWKV-7B [21]
- Llama-7B [22]

All these models were fine-tuned using the LoRA approach.
The fine-tuning parameters for each model are as follows: See Table 2:

Table 2. Fine-tuning Parameters For Each Major Model

	Chinese-Llama-7B	ChatGLM2-6B	RWKV-7B	LLaMa-7B
Iterations	3000	3000	3000	3000
Batch Size	1	1	1	1
Epoch	3	Dynamic [0.03,10.28]	[500, 100000]	3
Learning Rate	2e−4	2e−2	[4e−4, 1e−5]	2e−4
Network Layers	32	-	6	32
Embedding	1024	-	512	1024
Context Length	2048	128	1024	2048
β	0.9	-	0.9	0.9
ϵ	1e−8	1e−5	1e−8	1e−8

4.4   Experimental Environment

The experiments were conducted in the following two hardware environments:

- Single NVIDIA A100 GPU with 80G memory.
- Four NVIDIA A100 GPUs, each with 40G memory.

4.5   Experimental Process

4.5.1   Learning Domain Knowledge
Using LoRA technology, we fine-tuned the large models to help them learn the terms and business knowledge specific to the telecommunications domain. The domain knowledge fine-tuning dataset consisted of 10,000 entries. Some examples are provided below:

\什么是移动业务出账用户融合渗透率（宽移、移移、金融分期）？
指报告期末移动业务出账用户中宽移融合、移移融合、金融分期用户数
与移动业务出账用户数的比值。
其中：宽移融合、移移融合、金融分期用户数汇总剔重。

什么是移动业务赠款销账用户数？
指报告期内移动业务用户中有赠送话费优惠的用户数。

什么是联通看家生效套餐数？
指报告期内生效的联通看家业务（云存储服务）套餐量，
包含报告期内受理次月1日生效的套餐量。

主营业务收入同比增长率指什么？
报告期内主营业务收入较上年同期值增长率，计算公式：
（本期主营业务收入–上年同期主营业务收入）/上年同期主营业务收入。"

4.5.2   Task Completion - NL2SQL Fine-tuning was performed on the telecommunications dataset using instruction-based methods.

The format of the training sample is as follows:

```
{
"instruction": "现有一张名为DM_KPI_DATA_MARKET_M_ZNZS的SQL表,
其列名和对应中文解释如下:
Y_LY_VALUE: 上年同期累计值;
MONTH_ID: 账期(月);
Y_TY_VALUE: 本年累计值;
M_LY_VALUE: 上年同期值;
PROV_ID: 省分ID;
CITY_ID: 区县ID;
KPI_CODE: 指标编码;
KPI_NAME: 指标名称;
RJHB_VALUE: 日均环比;
LJTB_VALUE: 累计同比;
DATA_TIME: 具备时间;
PROV_DESC: 省分名;
AREA_DESC: 地市名; KPI_VALUE: 当月值; M_LM_VALUE: 上月值;
AREA_ID: 地市;
HB_VALUE: 环比。",

"input": "查询为2022年8月的海南省宽带接入未出账用户数的当月值、
上年同期值、本年累计值是多少? ",

"output": "select PROV_DESC,MONTH_ID,AREA_DESC,'宽带接入未出账用户
数',
KPI_VALUE,M_LY_VALUE,Y_TY_VALUE
from DM_KPI_DATA_MARKET_M_ZNZS where PROV_DESC='海南'
and MONTH_ID='202208' and AREA_DESC='汇总'
and CITY_ID='-1' and KPI_NAME='宽带接入未出账用户数'"
}
```

4.5.3   Result Validation

Automatic Validation. Initially, the SQL statements generated by the model were directly applied to the database tables for querying. If the output of the query is NULL, this implies the data doesn't exist, and we proceed to manual verification. If there's no output or if an error occurs, it indicates that the generated SQL statement is incorrect. If a numerical result is produced, it progresses to the manual verification phase.

Manual Verification. For the results produced by the model, manual table lookups were conducted to confirm the presence of the values and to ascertain the accuracy of the retrieved data. The verification process can be described as follows: See Fig. 4:

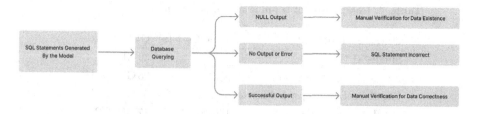

Fig. 4. Manual Verification process

Result Compilation. Based on the outcomes of the manual verification, the correctly generated SQL statements can be categorized into two groups: 1. Data Non-existent: This refers to the scenarios where the model correctly identified that the requested data does not exist in the database. 2. Correct Query Results: This pertains to instances where the model generated the right SQL statement, leading to accurate retrieval of existing data.

The accuracy rate can then be calculated using the following formula:

$$AccuracyRate$$
$$= \frac{\#ofSQLStatmentsGenerated}{\#ofAllOutputs}$$
$$= \frac{\#ofNULLOutputs + \#ofCorrectSQLStatmentsGenerated}{\#ofAllOutputs} \quad (4)$$

#: Number

This accuracy rate gives a clear indicator of how well the model performs in generating SQL queries based on natural language inputs. It indicates both the model's ability to recognize non-existent data and its capability to accurately retrieve existing data.

5   Experimental Result and Analysis

On the foundation of four open-source large models, we first conducted domain knowledge learning and then fine-tuned the task NL2SQL. The results of fine-tuning and inference are shown in Table 3

Table 3. Fine-tuning Parameters For Each Major Model. Data = 5000 Data from anonymous Internet Provider (AIP)

Data	Pre-trained Model	Accuracy on Train	Accuracy on Test
AIP	Chinese-Llama-7B	0.98	0.95
AIP	LLaMa-7B	0.93	0.7
AIP	ChatGLM2-6B	0.91	0.7
AIP	RWKV-7B	0.83	0.6

Table 4. Comparison of training methods and inferencing time cost

Distributed Method	Time Cost (s)	GPU Usage
Single GPU	178.40	39 GB
Parallelization:torchrun	55.20	38 GB/GPU
Parallelization:deepspeed-zero2	59.68	35 GB/GPU
Parallelization:Pipeline	196.09	17 GB/GPU

To facilitate the deployment of the fine-tuned large model for practical business, we used Chinese-Llama-7b as the base model and compared the duration required by different training and inference methods. The comparison results are shown in Table 3 and Table 4.

From the experiments, we can observe that the model built on Chinese-Llama-7b performs the best. The original LLaMA corpus is primarily in English, while Chinese-Llama-7b incorporated Chinese content and expanded the vocabulary, thus offering better support for the Chinese language. Although our experiment mainly focused on the quality of SQL generation without considering the framework and other factors, the results might be strongly correlated with the dataset. Nevertheless, this validates our initial intention: fine-tuning large models using the LoRA technique can accomplish vertical tasks (Fig. 5).

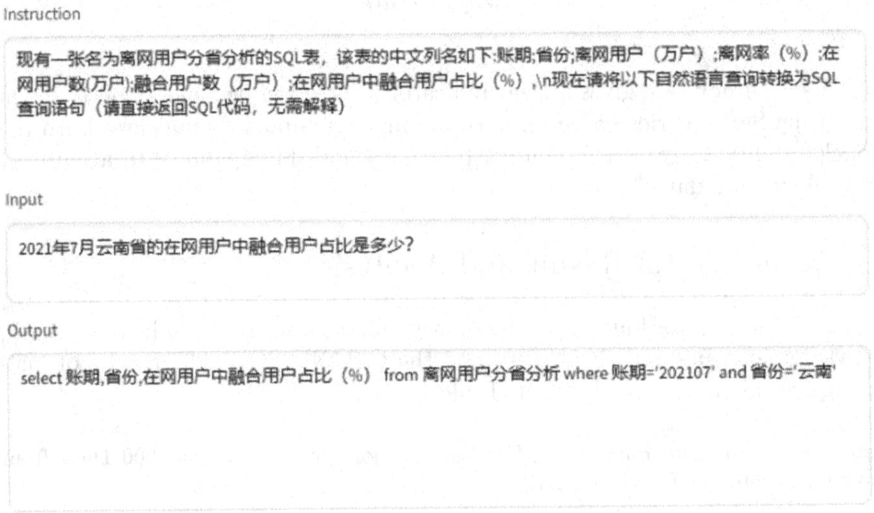

Instruction

现有一张名为离网用户分省分析的SQL表，该表的中文列名如下:账期;省份;离网用户（万户）;离网率（%）;在网用户数(万户);融合用户数（万户）;在网用户中融合用户占比（%）,\n现在请将以下自然语言查询转换为SQL查询语句（请直接返回SQL代码，无需解释）

Input

2021年7月云南省的在网用户中融合用户占比是多少？

Output

select 账期,省份,在网用户中融合用户占比（%） from 离网用户分省分析 where 账期='202107' and 省份='云南'

Fig. 5. A Sample of Inference Result of querying Yunan's Data

Table 5. Infrastructure of Large Model

Inference Method	Numerical Precision	GPU Usage	Time Cost(s)
Single GPU	FP16	28 GB	17.23
Single GPU	INT8	16 GB	49.43
Muti-GPUs*	FP16	8 GB/GPU	34.45

* On One-Node Pipeline

Through different training methods, see in Table 5, we can balance the utilization of time and space to meet our practical requirements. By employing various inference methods, we can choose the foundational infrastructure for the large model based on real-world situations.

6 Conclusion

The use of large-scale models, especially in vertical industries, demonstrates immense potential and challenges. In this study, using data from a specific operator as an example, we conducted experiments on the LoRA fine-tuning technique with 5000 sample data in a hardware environment of a single card A100 80G and four cards A100 40G. The performance of the fine-tuned model on the test dataset met expectations, reflecting strong capabilities. This confirms the efficacy of the fine-tuning strategy and the adaptability of the chosen model. Moreover, leveraging the text understanding ability of large models combined with LoRA tuning, we found that the training complexity, computational resource requirements, and storage resource needs of the NL2SQL model were significantly reduced. This provides a more economical and efficient solution for practical applications. It also exemplifies the empirical application of large models in vertical domains.

This research demonstrates that the model fine-tuning technique based on instruction fine-tuning can generate large models that meet specific vertical application requirements. By comparing different large model fine-tuning, training, and inference strategies, we validated the applicability of large models in specific vertical tasks and ensured that the costs of training and inference are manageable.

Fine-tuning based on instruction provides a pathway for developers with limited but high-quality data to explore their large model application needs. It also offers enterprises a concrete product implementation plan in vertical domains. Its controllable costs will also make the widespread application of large models feasible.

References

1. Radford, A., Narasimhan, K., Salimans, T., et al.: Improving language understanding by generative pre-training, pp. 8–25 (2018)
2. Liu, Y., Ott, M., Goyal, N., et al.: Roberta: A robustly optimized Bert pretraining approach, pp. 24 ArXiv preprint arXiv:1907.11692 (2019)
3. Narayanan, D., Shoeybi, M., Casper, J., et al.: Efficient large-scale language model training on GPU clusters using Megatron-lm. Proceedings of the International Conference for High-Performance Computing, Networking, Storage, and Analysis, pp. 35–39 (2021)
4. ChatGPT and Open-AI Models: A Preliminary Review. 15, 6, 192. https://doi.org/10.3390/fi15060192.ChatGPT.(n.d.). Will ChatGPT replace search engines?, p. 3
5. ChatGPT. (n.d.). Will ChatGPT replace search engines?
6. ChatGPT. (n.d.). Explaining some common misconceptions about large language models, pp. 8–25
7. Mitchell, E., Lee, Y., Khazatsky, A., et al.: DetectGPT: ZeroShot Machine-Generated Text Detection using Probability Curvature, p. 19. ArXiv preprint, abs/2301.11305 (2023)
8. Amatriain, X.: Transformer models: an introduction and catalog, pp. 8–41 (2023)
9. Korthikanti, V., Casper, J., Lym, S., et al.: Reducing activation recomputation in large transformer models, p. 39 ArXiv preprint arXiv:2205.05198 (2022)
10. Han, S., Mao, H., Dally, W.J.: Deep compression: Compressing deep neural networks with pruning, trained quantization, and Huffman coding, p. 45. ArXiv preprint, abs/1510.00149 (2015)
11. GPTs are GPTs: An Early Look at the Labor Market Impact Potential of Large Language Models. http://arxiv.org/abs/2303.10130 Accessed 05 Sep 2023
12. Yuan, S., Zhao, H., Du, Z., et al.: WuDaoCorpora: a super large-scale Chinese corpora for pre-training language models. AI Open **2**, 65–68 (2021)
13. Wei, J., Bosma, M., Zhao, V., et al.: Fine-tuned Language Models are Zero-Shot Learners. In: Proceedngs of ICLR, p. 54 (2022)
14. Sanh, V., Webson, A., Raffel, C., et al.: Multitask prompted training en- ables zero-shot task generalization. In: Proceedings of ICLR, pp. 54 (2022)
15. Hu, E.J., et al.: Lora: low-rank adaptation of large language models, p. 4. arXiv preprint hyperimagehttp://arxiv.org/abs/2106.09685arXiv:2106.09685 (2021)
16. Liu, X., et al.: P-tuning v2: Prompt tuning can be comparable to fine- tuning universally across scales and tasks, p. 2. arXiv preprint arXiv:2110.07602 (2021)
17. Liu, Y., Agarwal, S., Venkataraman, S.: Autofreeze: automatically freezing model blocks to accelerate fine-tuning. arXiv preprint arXiv:2102.01386. 18. (pp. 5) (2021)
18. Sun, X., Ji, Y., Ma, B., Li, X.: A Comparative Study between Full-Parameter and LoRA-basedFine-Tuning on Chinese Instruction Data for Instruction Following LargeLanguage Model, p. 2 (2023). https://arxiv.org/pdf/2304.08109.pdf
19. Ziqingyang/Chinese-llama-2-7b · hugging face. ziqingyang/chinese-llama-2-7b · Hugging Face. (n.d.). https://huggingface.co/ziqingyang/chinese-llama-2-7b
20. THUDM/CHATGLM2-6B · hugging face. THUDM/chatglm2-6b · Hugging Face. (n.d.). https://huggingface.co/THUDM/chatglm2-6b
21. Raven RWKV 7B - a hugging face space by blinkdl. Raven RWKV 7B - a Hugging Face Space by BlinkDL. (n.d.). https://huggingface.co/spaces/BlinkDL/RWKV-World-7B
22. Decapoda-research/llama-7b-HF · hugging face. decapoda-research/llama-7b-hf · Hugging Face. (n.d.). https://huggingface.co/decapoda-research/llama-7b-hf

Author Index

Printed in the United States
by Baker & Taylor Publisher Services